Teubner-Reihe UMWELT

B. Hock (Ed.)
Bioresponse-Linked Instrumental Analysis

Teubner-Reihe UMWELT

Herausgegeben von

Prof. Dr. mult. Dr. h.c. Müfit Bahadir, Braunschweig
Prof. Dr. Hans-Jürgen Collins, Braunschweig
Prof. Dr. Bertold Hock, Freising

Diese Buchreihe ist ein Forum für Veröffentlichungen zum gesamten Themenbereich Umwelt. Es erscheinen einführende Lehrbücher, Monographien und Forschungsberichte, die den aktuellen Stand der Wissenschaft wiedergeben.

Das inhaltliche Spektrum reicht von den naturwissenschaftlich-technischen Grundlagen über umwelttechnische Fragestellungen bis hin zu juristisch, sozial- und gesellschaftwissenschaftlich ausgerichteten Titeln. Besonderer Wert wird dabei auf eine allgemeinverständliche, dennoch exakte und präzise Darstellung gelegt. Jeder Band ist in sich abgeschlossen.

Die Autoren der Reihe wenden sich vorwiegend an Studierende, Lehrende sowie in der Praxis tätige Fachleute.

Bioresponse-Linked Instrumental Analysis

Edited by

Prof. Dr. Bertold Hock
Technische Universität München

 B.G.Teubner Stuttgart · Leipzig · Wiesbaden

Prof. Dr. Bertold Hock
Fax: + 498161714403
e-mail: hock@weihenstephan.de

Die Deutsche Bibliothek – CIP-Einheitsaufnahme
Ein Titeldatensatz für diese Publikation ist bei
Der Deutschen Bibliothek erhältlich.

1. Auflage Januar 2001

Alle Rechte vorbehalten
© B. G. Teubner GmbH, Stuttgart/Leipzig/Wiesbaden, 2001

Der Verlag Teubner ist ein Unternehmen der Fachverlagsgruppe BertelsmannSpringer.

www.teubner.de

Gedruckt auf säurefreiem Papier
Umschlaggestaltung: Peter Pfitz, Stuttgart

ISBN-13: 978-3-519-00316-8   e-ISBN-13: 978-3-322-86568-7
DOI: 10.1007/978-3-322-86568-7

Preface

Environmental monitoring requires new concepts, especially with respect to large scale screening programs. It appears to be most important to include warning systems that indicate toxicological or pharmacological effects on biosystems such as genotoxicity, neurotoxicity, immunotoxicity, cell toxicity or endocrine disrupting events. Therefore information should be obtained on the presence and concentrations of chemical species that individually or in combination may interfere with those essential life processes.

Analytical methods provided by chemistry have reached their limits if toxicological and pharmacological effects of chemical contaminants in samples are to be assessed. On the other hand, biological approaches are not capable to identify the chemical structure of environmental contaminants. As current approaches have limitations either on the analytical or on the bioresponse site, new ideas are required. A solution is seen in a skilful combination of bioassays and chemical analysis, which is defined as bioresponse-linked instrumental analysis. It hyphenates two processes, biomolecular recognition initiating a biological effect and chemical analysis.

The concept of bioresponse-linked instrumental analysis was born out of the collaboration between scientists from different disciplines: chemical analysis, bioanalysis and biosensors. The plans for this book originated from intensive discussions among experts, most of them members of the working group "Wirkungsbezogene Analytik" established in 1998 by the "Wasserchemische Gesellschaft" within the "Gesellschaft Deutscher Chemiker (GDCh)". It is hoped that this book will serve in the near future as a source for new approaches to environmental analysis, which are indispensable to cope with future demands. At the same time, it should be useful for students of chemistry and biology as well as experts interested in environmental analysis.

6

Emphasis has been laid in this book on biological components used as targets for active substances. It is acknowledged that there is a tremendous range of functions in living organisms that are susceptible to environmental contaminants, resulting in acute or chronic effects. However, key processes can be determined at the molecular and subcellular level which, after disruption, lead to a potential damage of highly complex target reactions and structures. The primary aim of bioresponse-linked analysis is the utilization of such key processes whose disruption becomes visible at the complex level of growth, development or reproduction. This holds true for both, human- as well as ecology-related analytics.

I would like to thank the Teubner-Verlag, especially Mr. Jürgen Weiß, for his cooperation and active support. I am grateful to Stefanie Rauchalles for her help editing this book.

Freising-Weihenstephan, July 2000   Bertold Hock

Contents

1 APPROACHES TO BIORESPONSE-LINKED INSTRUMENTAL ANALYSIS: ON-LINE COUPLING OF LIQUID CHROMATOGRAPHY TO BIOLOGICAL ASSAYS

D. van Elswijk[1], T. Schenk[2], U.R. Tjaden[2], J. van der Greef[2], H. Irth[3]

[1]Screentec BV, Niels-Bohrweg 11-13, 2333 CA Leiden, The Netherlands
[2]Division of Analytical Chemistry, Leiden/Amsterdam Center for Drug Research, Leiden University, P.O.Box 9502, 2300 RA Leiden, The Netherlands
[3]Vrije Universiteit Amsterdam, Department of Analytical Chemistry and Applied Spectroscopy, De Boelelaan 1083, 1081 HV Amsterdam, The Netherlands

Abstract. During the recent years on-line coupling of high performance liquid chromatography (HPLC) with bioassays, has evolved into a rapid, selective and above all a highly sensitive methodology, which is characterized by a broad range of interesting applications. In contrast to conventional microtiter type assays, the combination of separation power offered by LC and the high selectivity and sensitivity featured by bioassays enables the separate quantification of cross-reactive compounds. Moreover, on-line coupling of HPLC and bioassays overcomes both tedious fraction collection and manual operations, which are required to prepare fractions for batch biological assays, thus increasing the speed of analysis dramatically. This overview presents different strategies to implement biomolecular affinity interactions in on-line biochemical detection (BCD) techniques. Initially, on-line continuous-flow BCD-systems were primarily focussed on low molecular mass targets. Recently, however, we developed BCD strategies suitable for detecting high molecular mass ligands (e.g., proteins) as well, using either labeled receptors or labeled antibodies.

1.1   Introduction

During the last decade trends can be seen towards the implementation of biochemical methods in conventional analytical systems. The main objective of this approach is to combine the selectivity of biospecific interactions, i.e., receptor-ligand, enzyme-inhibitor, antibody-antigen interactions with the separation, identification power and ease of automation of modern analytical techniques such as liquid chromatography-mass spectrometry. This approach leads to analytical methods, which simultaneously provide chemical and biochemical information. The main application area lies in the detection and discovery of biologically active substances in complex matrices, ranging from drug discovery (e.g., natural product screening) to environmental screening.

The most common way to implement biospecific interactions in analytical methodologies is based on the immobilization of the affinity protein, mostly an antibody against the analyte, onto a solid support. Immunoaffinity supports have been exploited in the sample handling area as a highly selective, immunopreconcentration step (de Frutos et al. 1993). Though substantially increasing the selectivity of the method, a similar enhancement of sensitivity, however, is generally not obtained. Immunoassays, on the other hand, are generally characterized by an improvement of detection sensitivity by two or three orders of magnitude (Gosling 1990). By employing radioactive, fluorescent or enzyme labels, the intrinsic detection properties of the analyte are greatly enhanced, thus decreasing detectable concentrations to impressive limits. However, despite the gain in sensitivity the major limitation of bioassays in general emanates from the fact that the interacting affinity biomolecules, such as antibodies or receptors, could exhibit cross-reactivity to structurally related compounds, which would strongly interfere with the detection of the parent compound, leading to erroneous results. Consequently, in order to overcome the problems associated with cross-reactive compounds, separation methods such as HPLC have been employed frequently as a fractionation step prior to the immunoassay (Gelpi et al. 1989, Vetticaden et al. 1986, Harajiri et al. 1992, Gelpi 1985). The advent of flow-injection immunoassays, originally developed as an alternative for microtiter-type based assays, allowed gross automation, improving assay times considerably. Flow-injection assay times below 10 min have been reported frequently (Gübitz 1990).

(A)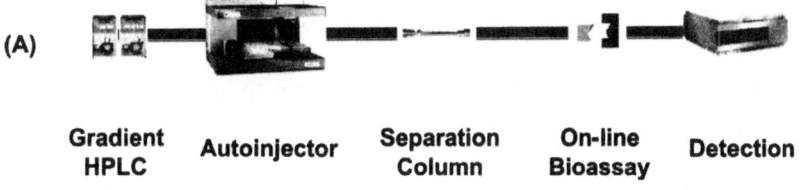

| **Gradient HPLC** | **Autoinjector** | **Separation Column** | **On-line Bioassay** | **Detection** |

(B)

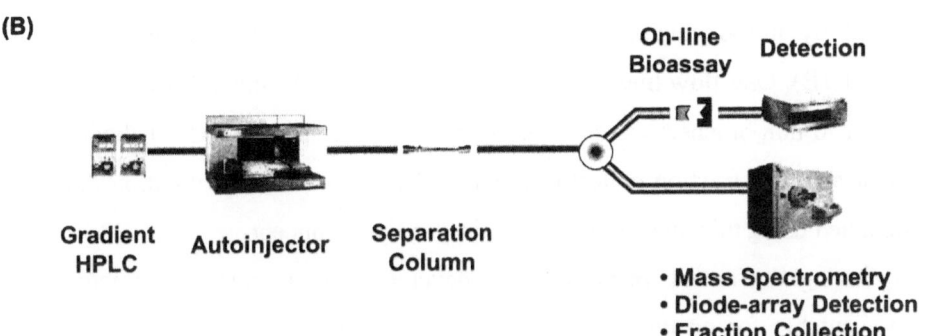

On-line Bioassay   **Detection**

Gradient HPLC   **Autoinjector**   **Separation Column**

• **Mass Spectrometry**
• **Diode-array Detection**
• **Fraction Collection**

Fig. 1.1: A, Principle of of-line liquid chromatography biochemical detection, B, simultaneous LC-BCD and mass spectrometry

Despite the significant decrease of assay times, i.e. minutes rather than hours, flow-assays are mostly carried out sequentially, simulating reagent addition, washing and incubation steps, similar to those performed in microtiter plates (Shellum and Gübitz 1989, Locascio-Brown et al. 1988, Wittmann and Schmid 1993, Kramer and Schmid 1991a, b, Evans et al. 1994, Palmer et al. 1993, Wortberg et al. 1994,

Gunaratna and Wilson 1993, Liu et al. 1991, Tang et al. 1991, Mattiasson et al. 1990, Arefyev et al. 1990, Cassidy et al. 1992, Kusterbeck et al. 1990). Consequently, sequential addition immunoassays (SAIA) performed in flow injection systems do not allow the continuous monitoring of the LC effluent, making these types of assays unsuitable for on-line coupling to LC. On the other hand, a number of truly continuous flow immunoassays, capable of continuously measuring immunoaffinity interactions, have been reported (Freytag et al. 1984, Whelan et al. 1993).

In the present paper we are reviewing our efforts to develop integrated biochemical detection (BCD) methods, i.e, analytical systems where the biological assay is an integral part of an HPLC-based screening method. Figure 1.1A illustrates the concept of post-column biochemical detection, the biochemical assay being performed on-line to an HPLC separation. This basic concept can be extended by introducing a flow-split directly after the LC separation column (Figure 1.1B). One flow line is directed towards the biochemical detection system, the other to non-biochemical detectors such as a mass spectrometer or diode-array detector and/or a fraction collector. Figure 1.2 illustrates the information that can be obtained using this integrated biochemical analysis concept.

During the recent years we have developed continuous-flow biochemical assays suitable for on-line coupling to liquid chromatography. The coupling of separation power of LC on the one hand and the selectivity and sensitivity offered by bioassays on the other hand, enables the rapid analysis of cross-reactive compounds at low concentration levels. Moreover, the traditional laborious manual operations and rather long assay times, characteristic for microtiter plate-based assays, are jointly overcome. Additionally, the possibility to perform sample clean-up, using both pre-column sample handling as well as the chromatographic process itself, represents another potential advantage of on-line systems. By eliminating high ionic strength and low- and high-molecular weight matrix components, biochemical detection is carried out in a rather "clean" and well-defined environment, which results in improved reproducibility.

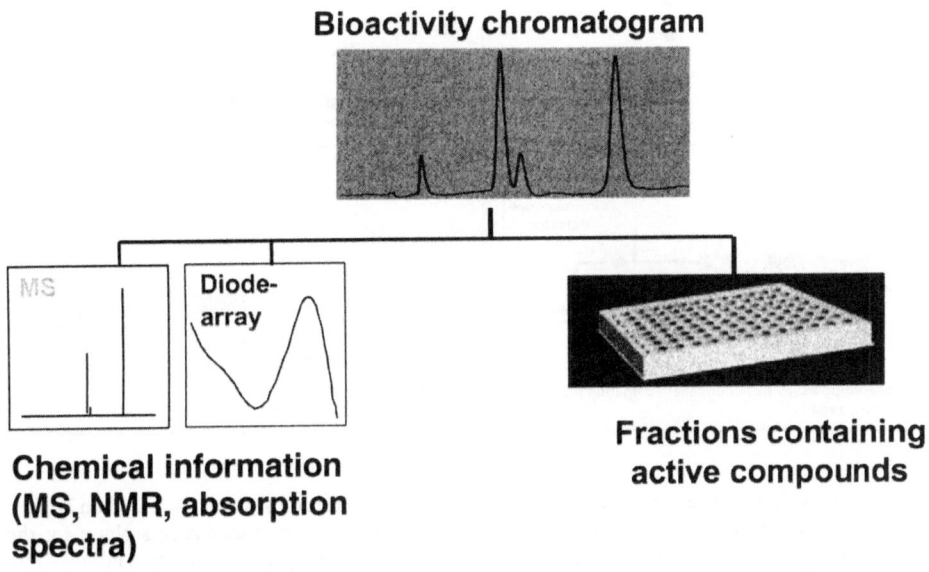

Fig. 1.2: Information obtained from integrated biochemical detection systems

The ability to continuously monitor the LC effluent selectively for low concentrations of known or unknown affinity analytes, makes HPLC-BCD especially suited for applications in the area of bioanalysis and drug discovery. This article reviews the efforts made to couple HPLC on-line to various types of continuous-flow biochemical assays. Apart from BCD techniques capable of detecting small molecular compounds, like digoxin and leukotrienes, a new line of applications concerning the bioanalysis of protein biomarkers (Oosterkamp et al. 1998, Schenk et al., submitted) such as several cytokines is discussed. By describing these setups, detection of both small molecular compounds, like estrogens, as well as high molecular mass compounds, such as proteins, is illustrated.

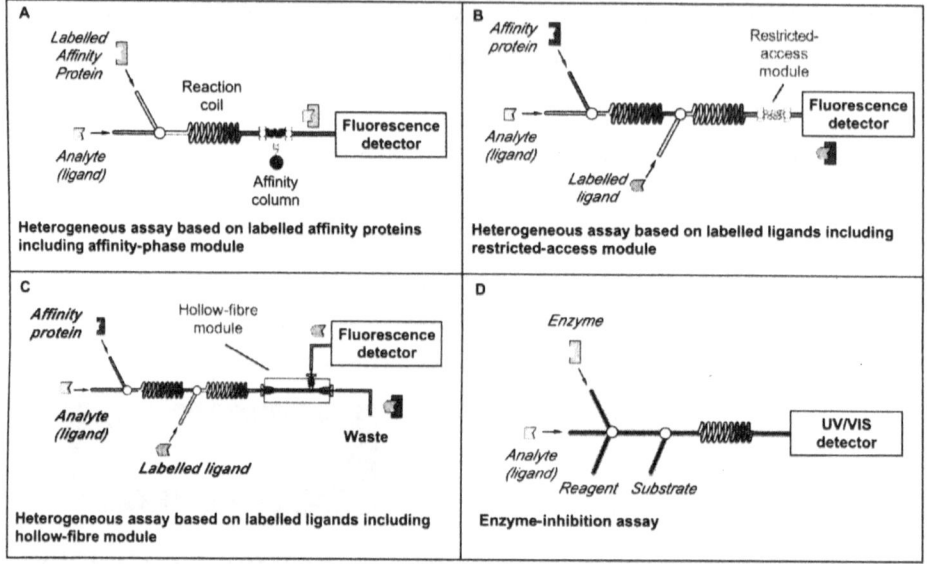

Fig. 1.3: Assay designs used on continuous-flow biochemical detection systems. A, based on labeled affinity proteins, B, based on labeled ligands and using a restricted-access phase for the separation of free and bound fraction, C, based on labeled ligands and using a hollow-fibre membrane for the separation of free and bound fraction, D, homogenous enzyme-inhibition assay

1.2   Biochemical detection strategies for the determination of low molecular weight compounds

1.2.1   Labeled antibodies as reporter molecules

The first BCD-strategy we developed, is based on soluble labeled affinity proteins, such as antibodies or receptors (Figure 1.3A) and enables the quantification of a wide range of compounds, ranging from small molecular weight molecules, such as digoxin and leukotrienes, to high molecular weight compounds including urokinase and cytokines. After chromatographic separation of the antigenic analytes, labeled affinity proteins are added to the LC effluent. During a certain period of time, usually in the order of one min, the affinity proteins are allowed to react with the antigenic compounds, eluting from the LC column. After complex formation, the excess of unreacted labeled affinity protein is trapped on a short affinity column packed with an immobilized antigen support. As labeled affinity protein/analyte complexes are not retained by the affinity column, detection of these complexes by conventional LC fluorescence detectors is allowed. The concentration of labeled affinity protein/analyte complex formed during the affinity interaction depends on several parameters. Most important features, however, are the reaction times themselves and the concentrations of both analyte(s) and affinity proteins . During the recent years we have developed on-line LC-BCD techniques, which used both labeled affinity proteins (Irth et al. 1992, Oosterkamp et al. 1994a) and labeled ligands (Oosterkamp et al. 1994b, 1996a, b, Lutz et al. 1996) as reporter molecules. For the detection of small molecular weight molecules, using soluble affinity proteins, the choice of label type has often been dictated by the sensitivity of the BCD strategies involved. However, depending on the type of application, separation of free and bound labels, is not always straightforward and often requires creative solutions, which may narrow the range of usable reporter molecules In order to demonstrate the HPLC-BCD strategy, digoxin and its cross-reactive metabolite digoxigenin are used as model compounds (Irth et al. 1992, Oosterkamp 1994a). Separation of digoxin and its cross-reactive metabolite digoxigenin is achieved by reversed phase HPLC using an analytical column packed with C18-silica. The reversed phase LC separation is performed under isocratic conditions, employing 30% of acetonitrile in the mobile phase. Subsequently the LC effluent is diluted threefold by post-column addition of bioreagent, thus reducing the amount of organic modifier to 10%. Using this BCD-strategy, digoxin and digoxigenin could be detected down to 4 and 1 nM concentration levels (100_μl injections) in both plasma and urine. Figure 1.4 demonstrates the high selectivity and sensitivity obtained when detecting both compounds in urine (100 nM injections). The fluorescence response close to the void volume region, possibly originates from cross-reactive steroid compounds, naturally present in urine. However, this clearly illustrates the advantage of on-line BCD compared to microtiter batch assays, i.e., enabling the selective monitoring of cross-reactive compounds in a particular sample instead of obtaining one summarized fluorescence signal corresponding to several cross-reactive compounds. In addition, this example also demonstrates the potential of drug discovery using LC-BCD. As a consequence of these highly selective and sensitive detection methods, sample pretreatment requirements are less strict for most biochemical detection systems. HPLC-BCD set-ups capable of directly injecting plasma or urine samples, using efficient solid phase extraction on C18 restricted access columns, have recently been reported (Oosterkamp et al. 1996a). In addition, the utilization of labeled antibodies, or affinity proteins in general, is particularly useful when batches of high purity, containing either labeled or unlabeled affinity proteins, are commercially available. The presence of non-specific or inactive affinity proteins or other labeled proteins, is likely to increase the fluorescent background as they are not retained by the affinity column. Consequently, sensitivity highly depends on the purity of the labeled protein preparation.

Considering the fact that usually only unpurified antiserum or unlabeled poly- or monoclonal antibodies are commercially available, we investigated the development of biochemical detection systems, which employ untreated antisera or antibody preparations using labeled antigens as reporter molecules.

1.2.2   Labeled antigens as reporter molecules

As has been described previously, the biochemical detection strategy employing labeled antibodies is based on direct detection of the analyte, i.e., both analyte recognition as well as quantification occur through the labeled antibodies themselves. In contrast, antibodies in "labeled antigen setups" are merely used for affinity recognition of analytes, whereas quantification is carried out indirectly using labeled antigens. Typically, LC separation of the analytes is followed by post-column addition of an unlabeled affinity protein preparation. The mixture is allowed to react for a short period of time, after which a preparation of labeled antigens is added to the mobile phase. The excess of binding sites left, depending on the analyte concentration in the sample, is titrated via the reporter molecules. Both the concentration of labeled antigens left as well as the concentration of labeled complexes formed are a measure for the original analyte concentration.

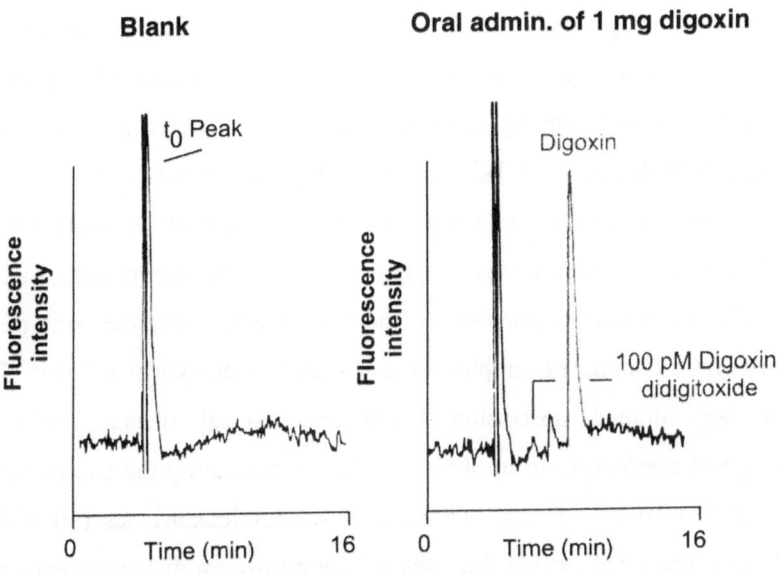

Fig. 1.4: LC-BCD of blank urine (left) and urine (right) spiked with (1) 10^{-8} M digoxigenin and (2) 10^{-8} M digoxin (100 μl injections)

At this point several biochemical detection strategies can be employed in order to actually quantify the amount of analytes. Figure 1.3B displays a frequently applied BCD strategy, which is particularly suitable for the detection of small molecular weight compounds. After addition of both affinity protein and labeled antigen, complexes of affinity protein-labeled antigen are subsequently separated from excess unreacted labeled antigen by means of a restricted access column. The hydrophilic particle surface combined with small hydrophobic pores (app.10kDa) enables the passage of both affinity protein and affinity complex as their bulky nature (150 kDa) prohibits entering the small pores, whereas unreacted labeled antigen is efficiently retained. Depending on the conditions chosen, a certain percentage of the labeled antigens binds to the affinity proteins, which results in stable fluorescent background signal. Injecting a certain concentration of analyte results in a decrease of binding sites available for the labeled antigens. Consequently, during a short period of time, the amount of labeled antigens trapped is increased, whereas the concentration of labeled antigen/affinity protein is temporarily decreased, causing a negative signal in the fluorescence background.

Several leukotrienes, LTC4, LTD4 and LTE4 respectively, were used as model compounds in order to demonstrate this BCD strategy. On-line SPE for the direct injection of protein containing samples was performed using a Lichrospher RP-4 ADS precolumn. Separation of the cross-reactive analytes was performed under isocratic conditions, employing a C4 packed analytical column. The LC-effluent was diluted post-column with biochemical reagent, reducing the percentage of acetontirile from 30% to 10%. By comparing the analysis of a non-stimulated human cell extract spiked with several leukotrienes (10 nM) and a stimulated human cell extract, the metabolization pattern of leukotrienes could be elucidated (Figure 1.5 A, B).

In contrast to microtiter plate assays, the ability to selectively detect reactive compounds during a single run enables LC-BCD to obtain a metabolic fingerprint, provided the affinity of the individual compounds is sufficiently high to allow detection. Apart from physically separating free and antibody-bound label in order to be able to quantify analyte concentrations, a feature characteristic for heterogeneous assays, a complementary strategy has been employed, which utilizes differences in labeled antigen detection properties upon binding to affinity proteins (quantum yield, excitation/emission wavelength).

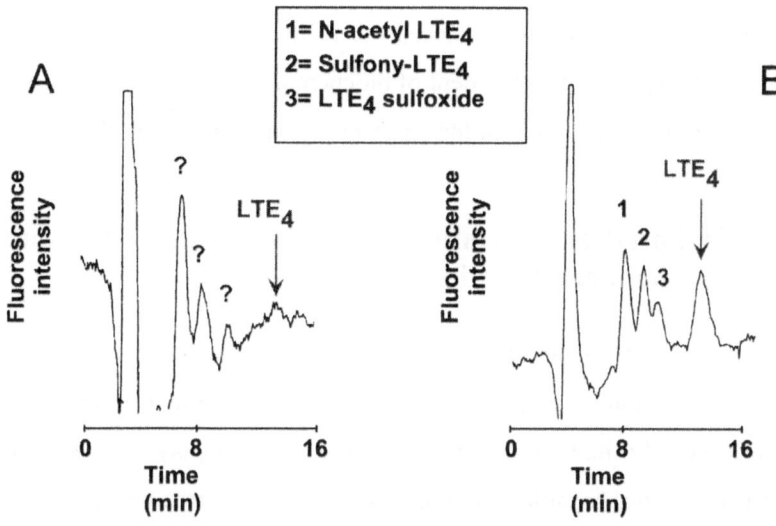

Fig. 1.5: LC-BCD of a human cell extract. A, blank, B, spiked with 10 nmol/l leukotrienes.1.LTC4, 2.LTD4, 3. LTE4; (500 µl injections), (c) extract from stimulated human cell culture

This BCD strategy, which allows biochemical detection without separation of free and bound label, a so-called homogeneous assay, has been chosen for the determination of sulfidopeptide leukotrienes in urine samples using BODIPY®-labeled leukotriene E4 as a label (Oosterkamp et al. 1996a). When the label was bound to the antibody, a significant difference in fluorescence quantum yield was observed, thus enabling direct detection of sulfidoleukotrienes and their cross-reactive metabolites. LTE4 in urine samples, for example, could be determined down to 0.2 ng/ml (2 ml injection). Apart from the quantification of known analytes, HPLC-BCD also proved to be useful in the discovery of cross-reactive LTE4 metabolites. On the other hand when free and bound-antibody label exhibit similar detection properties, a homogeneous setup will not allow quantification of the biochemical interactions, thus necessitating the separation of free and bound label. Trapping small hydrophobic molecules within the pores of restricted access material and allowing the subsequent detection of bound labeled antigens, certainly is an elegant detection method (Figure 1.3B). However, considering the fact that several labels, such as enzymes, exceed the restricted access pore diameter (>10 kD), we investigated the development of biochemical detection systems employing hollow-fibers as the principal tool for separation of free and bound labels.

1.2.3   Hollow-fibre membrane separation in continuous-flow biochemical detection

The limitations met so far in biochemical detection using labeled antigens are caused by the demands of the separation of free and bound label. By applying membranes these limitations may be overcome. For this purpose, we developed a hollow-fiber module, capable of continuously separating free and bound label in a flow system, due to their characteristic and considerable difference in size (Lutz et al. 1996). Free label, with a molecular mass generally below 1 kDa, freely passes the membrane, whereas antibody-bound label, with a molecular mass around 150 kDa, does not (scheme, see Figure 1.3C). By choosing a membrane with a molecular mass cut-off in between, a permeate stream only containing free label is obtained. In contrast to dialysis, the driving force of the separation is based on a pressure drop over the membrane. In this way, the separation of high and low molecular mass compounds is rapid and concentration-independent. For example, a hollow fiber with a molecular mass cut-off of 50 kDa allows the separation of antibodies from free molecular mass labels. A detection limit of 8 nmol/l was reached for the model compound biotin, employing anti-biotin antibodies and fluorescein-biotin as low molecular mass label. The advantage of LC-BCD for this model compound in comparison to LC-UV is clearly demonstrated in Figure 1.6.

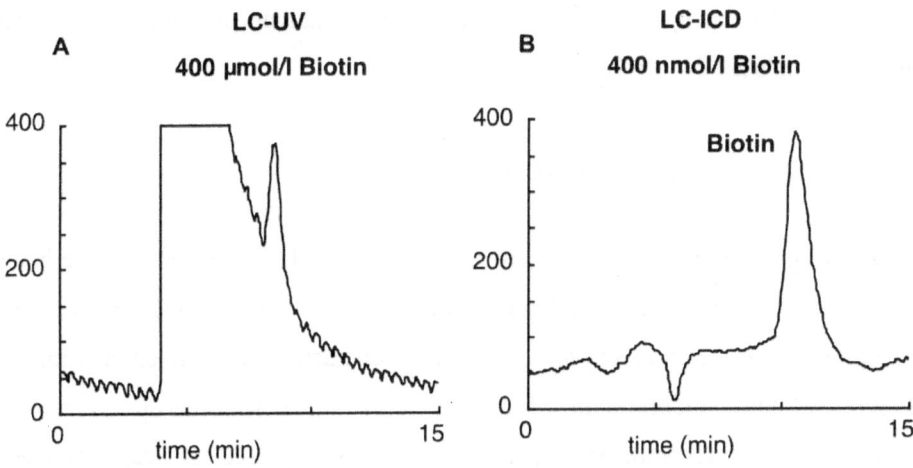

Fig. 1.6: A, LC-UV detection at 210 nm of 400 µmol/l biotin, B, LC-BCD of 400 nmol/l biotin

Selectivity is strongly increased and the signal to noise ratio is drastically improved. The front in the left chromatogram as well as the dip at t=6 min in the right chromatogram are caused by DMSO used to dissolve biotin. In contrast to restricted access columns, hollow-fibers allow the use of small-enzyme labels. Horseradish peroxidase (44 kDa), for example, may be separated from antibodies by a hollow-fiber with a cut-off of 100 kDa, thus allowing the implementation of commercially available enzyme bioassay kits in continuous-flow biochemical detection. Another major advantage of BCD setups employing hollow-fibers, is the possibility to introduce affinity solid phases, such as membrane-bound receptors, affinity beads or even cells, rather than the often used soluble affinity proteins. Recently, affinity solid phases, i.e. microscopic particles containing immobilized affinity proteins, have been used as affinity entities. Biotin, which acted as a model compound, could be detected down to 8 nmol/l, employing streptavidin coated solid-phases and fluorescein-biotin as low molecular mass label (Lutz et al. 1997). A similar setup should allow the analysis of solid-phase immobilized analytes using soluble affinity proteins for recognition and soluble labels for quantification.

This analytical challenge is found in the area of combinatorial chemistry where potential ligands are synthesized on a solid support.

1.3   Receptor affinity detection

During the recent years a series of on-line biochemical detection setups have been developed, which allow the implementation of basically any affinity interaction exhibiting high association and low dissociation characteristics. Implementation of receptors, whether soluble or membrane-bound, would present the opportunity to translate microtiter receptor assays, which are extensively used for drug discovery purposes, into continuous-flow biochemical detection setups. Similar to biological assays, quantification of the analyte is only possible, when either the receptor or a ligand act as a reporter molecule. However, the use of labeled receptors highly depends on the purity of the preparation. Ideally, some ligands have been reported to exhibit native fluorescence, making fluorescent labeling and the potential loss of affinity redundant (Nelson et al. 1984, Lee et al. 1977). Nevertheless, since native fluorescent ligands are scarce, fluorescence labeled ligands have been developed for a variety of receptors (Heithier et al. 1994). Recently our group developed two receptor affinity detection (RAD) setups, applicable to both low and high molecular weight compounds. The first RAD setup demonstrated the quantification of estrogens using a recombinant steroid binding domain of the human estrogen receptor as affinity protein and coumestrol, a fluorescent estrogen, as reporter molecule (Oosterkamp et al. 1996b). Using 5 nmol/l of human estrogen receptor, a detection limit of 5 nmol/l was obtained for compounds possessing a high affinity for the estrogen receptor such as 17β-estradiol and diethylstilbestrol. Weaker ligands, such as estriol and zeranol, displayed detection limits around 20 nmol/l. Non-estrogenic steroids such as methyltestosterone or progesterone were not detected at all.

In contrast to the model on-line receptor assay for small molecules based on the estrogen receptor we also developed a RAD for high molecular weight targets, such as proteins, using a labeled receptor (Irth et al. 1992). Urokinase plasminogen activator (uPA), which is involved in the proteolysis of biologically active peptides, was used as model protein (Ploug et al. 1992). The RAD system is based on the soluble receptor of uPA (uPAR; Behrendt et al. 1993, Kuiper et al. 1992). The principle of this assay is depicted in Figure 1.3A. A solution of fluorescent-labeled uPAR is added continuously to the LC effluent to react with uPAR-binding ligands. After a reaction time of 60 s, the excess of unreacted labeled uPAR is separated from the bound uPAR-ligand complex by means of a short affinity column packed with an immobilized uPA support. The uPAR-uPA complex passes the affinity column unretained and is detected downstream by means of a conventional fluorescence detector. Despite the low percentage of active labeled receptor, i.e. only 10% is able to bind to the immobilized uPA affinity column, still detection limits of 40 fmol and 800 fmol were obtained during flow injection analysis and on-line HPLC-RAD, respectively. Affinity purification of the labeled receptor preparation could strongly decrease the high fluorescent background caused by the high percentage of labeled receptor passing the affinity column unretained, thus increasing sensitivity even more.

Again, the advantage of coupling HPLC and BCD on-line, is demonstrated during uPA analysis using HPLC-RAD. A post column split was used to obtain a flow of 40 μl/min into the BCD system and 360 μl/min to an UV detector. To dilute the organic modifier gradient, the FLUOS-uPAR was added with 500 μl/min to the 40 μl/min eluting from the reversed phase separation. A typical chromatogram of the determination of uPA is shown in Figure 1.7, which depicts both the RAD and UV trace. UV analysis of standard uPA samples revealed a small peak eluting prior to the uPA peak. It possibly represents a breakdown product. In the RAD chromatogram the breakdown product results in a peak of almost equal height as the uPA peak, indicating that the breakdown product has a high affinity for the uPAR. In a batch receptor assay the low amount of uPA-breakdown product (small peak in UV trace) would, due to its high affinity, give rise to a strong response. The batch receptor assay would not distinguish between both active ligands present in the sample. In drug discovery the activity of this sample would probably have been designated solely to the high amount of low affinity compound and the trace amount of high affinity ligand would probably be overlooked and hereby missing a potential candidate drug.

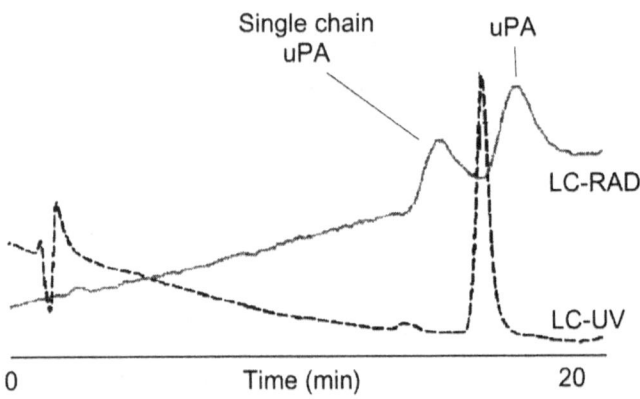

Fig. 1.7: Reversed phase chromatogram of 100 nmol/l uPA and its breakdown product (20 µl injection) using on-line receptor affinity detection

When analyzing human plasma, no background signal could be seen in the RAD fluorescence trace. uPA and its conversion product could readily be detected in plasma at concentrations of 16 µg/ml while no signal at all was obtained in UV-absorbance detection. This clearly demonstrates the selectivity of on-line HPLC-RAD in bioanalysis.

The on-line coupling of HPLC with receptor assays allows the determination of analytes based on biological activity. Cross-reactive compounds can be distinguished. Thus, the present approach can be a useful tool, for example, in metabolite research or drug discovery. The on-line approach provides rather short analysis times in the order of 15 to 20 min and overcomes laborious fractionation and evaporation steps. Due to the high selectivity of the detection method, sample handling is limited to a minimum, often only centrifugation of particulate matter is required. A post-column split provides simultaneous biochemical detection and, for example, UV, diode-array or mass spectrometric detection. In this way, a single analysis provides information on the distribution of biological activity in a particular sample, as well as providing structural data of active compounds. The most important application areas will clearly be drug discovery, particularly in the screening of complex samples such as natural product extracts. A second field of application of RAD would be metabolic profiling, as both parent drug and (active) metabolites can be determined in complex biological matrices.

1.4 On-line liquid chromatography-immunochemical detection of proteins

Recent developments in molecular biology have increased the availability of recombinant proteins. This has opened up the possibility to investigate the potentials of proteins as therapeutics or to use proteins as biomarkers to monitor physiological or pathological processes in the human body. Combining the high separation efficiencies and speed of HPLC with the extreme sensitivity of biochemical detection (LC-BCD) has already proven to be a very powerful and fast analytical tool for small, non-protein molecules. However, one of the main requirements for on-line coupling of both techniques is fast reaction kinetics (within minutes) which could be problematic when it comes to interactions between two macromolecular proteins. One could expect association reactions to be too slow for a post-column detection system for proteins due to the low diffusion rate of large proteins. Only recently, Miller and Herman (1996) presented the first on-line LC-BCD system for a protein, human methionyl granulocyte colony stimulating factor (GCSF), using both labeled antibodies and receptors (Newton et al. 1989).

Cytokines are group of (glyco)proteins of 10-30 kDa which play an important role in regulating immune responses and have been associated with several inflammatory diseases (Whicher and Evans 1990, Ingkaninan et al. 2000). As a consequence, cytokines are important biomarkers to monitor the effect of drugs influencing the immune system or inflammation. Problems often encountered in batch bioassays for cytokines like cross-reactive compounds, soluble cytokine receptors and oligomer formation could be overcome by on-line LC-BCD.

We developed an ICD system for cytokines analogous to the ICD system for small analytes, e.g., digoxin developed by Irth et al. (1992) (Figure 1.3A). The post-column ICD system presented here is based on the fast association of fluorescent-labeled antibodies with their target cytokine. In the first step, the antibodies are added to the mobile phase. Cytokines reacts with the antibodies to form fluorescent biocomplexes. In a second step free antibodies are removed prior to fluorescence detection via passage through a small column packed with a cytokine-bound support. After labelling of the anti-cytokine antibodies more than 90% of the fluorescently labeled antibodies was able to bind to the immobilised cytokine support. This high binding percentage resulted in a low fluorescence background and a large dynamic range. This limited fluorescence noise and high response resulted in detection limits, in flow injection, down to 2 fmol for Interleukine 4 (IL-4).

The influence of the reaction times in the coil and column on the performance of the system was investigated to get a better understanding of the reaction kinetics of macromolecules in postcolumn reaction detection techniques. Similar to ICD systems for small non-protein analytes developed earlier, reaction times were in the order of 1 minute. We also investigated the impact of different LC conditions like ionic strengths and organic modifiers on the performance of the ICD system. After unsuccessful attempts to couple the present ICD system to ion-exchange and size-exclusion chromatography we were able to couple post-column detection systems for interleukine 4, 6, 8 and 10 to micro-reversed phase chromatography.

However, mobile phases used in reversed phase HPLC such as methanol, ethanol or acetonitrile are capable of changing the tertiary structure of proteins, which might cause loss of their biological activity. For this reason severe problems could be expected when coupling reversed phase HPLC to ICD. However, the denaturing effect of the most common modifiers is reversible. If the modifier concentration decreases, the proteins will return to their original conformation. This offers the possibility to couple reversed phase separations to ICD, namely by diluting the HPLC effluent with the bioreagent to a modifier concentration which is not harmful to the protein. Narrow-bore LC was chosen as the separation step prior to the ICD. Namely, because narrow-bore LC offers the possibility to maintain the low flow rates in the ICD system (minimizing the reagent consumption) without post-column splitting. Narrow-bore LC also offers the possibility to dilute the HPLC effluent (40 μl/min) with the bioreagent solution (160 μl/min) and hereby decreasing the adverse effects of the salt or modifier gradients on the ICD system. A third advantage using narrow-bore LC is the enhanced pre-concentration effect when gradient separations are employed compared to normal bore LC.

Figure 1.8 shows chromatograms of 100 nM interleukine 6 (IL-6) and 100 nM interleukine 8 (IL-8) and a reversed phase separation between 100 nM IL-4 and 10 nM IL-10 coupled on line to biochemical detection using a methanol gradient. As can be seen the organic modifier gradient does increases the baseline of the ICD system. Experiments proved that this base-line increase was caused by the fluorescence enhancement of the background fluorescence, which does not bind to the cytokine affinity column. The fluorescence intensity of fluorescein strongly depends upon its environment. Organic modifiers are known to strongly increase the fluorescence intensity of fluorescein. Using a different fluorescent label of which the fluorescence intensity is not influenced by the organic modifier concentration will result in a stable baseline and thus a more sensitive LC-ICD system for cytokines.

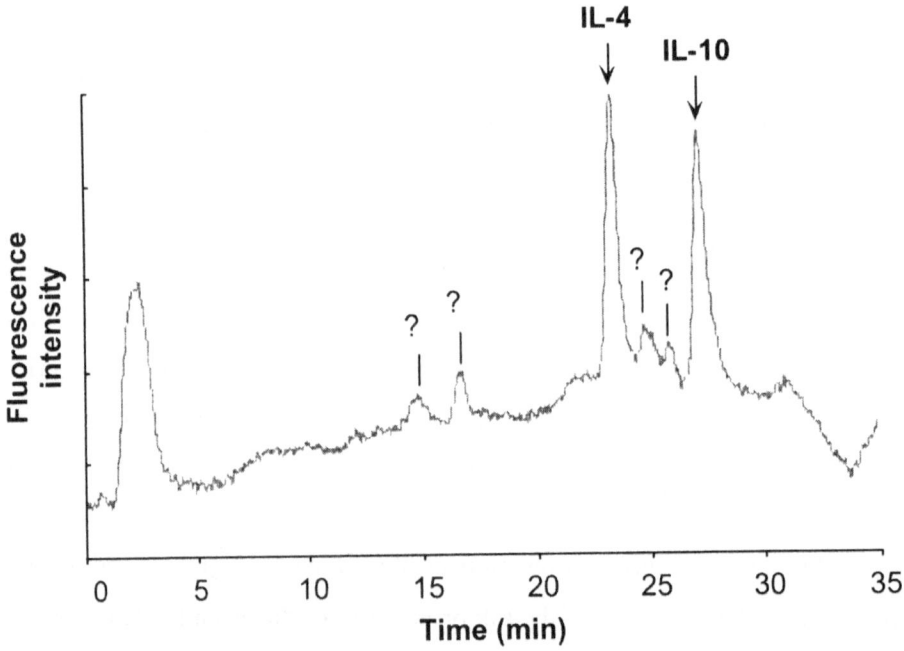

Fig. 1.8: Biochemical detection coupled on-line to gradient micro reversed phase chromatography. Separation between IL-4 (100 nM) and IL-10 (10 nM)

For the LC-ICD analysis of two cytokines in a single run the corresponding cytokine columns were placed in series. Attempts to co-immobilise two or more cytokines on one support were not successful. This limits the amount of different cytokines that can be detected in a single LC-ICD system because of the bandbroadening and increased backpressure when more than two cytokine affinity columns are placed in series. The cytokine samples used to obtain the chromatograms shown in Figure 1.8, contained 1 mg/ml blocking reagent e.g. a 10.000 times excess of other proteins. This also proves the extreme selectivity of the developed ICD system. Identical FI-ICD system as presented for IL-4, IL-6, IL-8 and IL-10 were also developed for other cytokines like tumor necrosis factor-α, Interferon-γ, Interleukine 2 and Interleukine 12. These cytokines could successfully be immobilised on a solid support. The antibodies to the cytokines could also be labeled while still binding to the immobilized cytokine column. However, when the cytokines were injected into the FI-ICD system only very small peaks appeared, resulting in very high detection limits of 100 nM. Different approaches to solve this problem have been unsuccessful so far.

1.5 Simultaneous liquid chromatography-biochemical detection and mass spectrometry

An interesting application area of integrated biochemical detection method is the discovery of unknown biologically active substance in complex matrices. An example is natural product screening where active compounds have to be detected and identified in complex samples containing many nonactive, low- and high-molecular mass substances at various concentration levels. LC-BCD systems can be employed in the separation of these compound mixtures; valuable information on the active compounds can be obtained when mass spectrometry is employed in parallel and simultaneous to the biochemical measurements (see Figure 1.1B).

This approach is exemplified by the screening of *Narcissus* extracts for acetylcholinesterase (AChE) inhibitors which may play an important role in the treatment of Alzheimer's disease. For this purpose a microtiter plate assay based on the Ellman reaction was converted into a homogeneous flow assay (Ingkaninan et al. 2000). This assay (scheme, see Figure 1.3D) is based on the enzymatic

30

hydrolysis of acetylthiocholine iodide to yield thiocholine. When thiocholine reacts with 5,5'-dithiobis-2-nitrobenzoate, it will produce 5-thio-2-nitrobenzoic acid, which has a strong UV/VIS absorption at 405 nm. AChE inhibitors prevent the

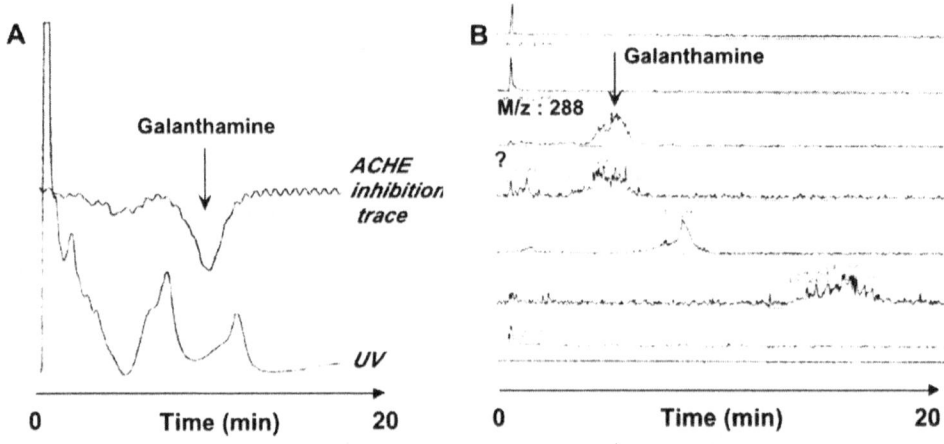

Fig. 1.9: LC-BCD-MS of a *Narcissus* extract using an acetycholinesterase inhibition assay. (a) LC-BCD (upper trace) and parallel UV (215 nm, lower trace). (b) Selected m/z traces recorded in series to the UV detector by electrospray MS

formation of 5-thio-2-nitrobenzoic acid resulting in a decrease of the signal which is monitored by a conventional VIS detector. By introducing a flow split after the LC column, the major part of the LC effluent was directed to a UV and subsequently to an electrospray MS detector. This allows the simultaneous monitoring of biochemical activity and recording of mass spectra. Due to a fixed time difference between the biochemical line and the MS line, peaks appearing in the BCD chromatogram can easily be correlated with the corresponding peaks in the MS.

A typical chromatogram of a *Narcissus* extract is shown in Figure 1.9. The two peaks in the BCD chromatogram indicate the presence of AChE inhibitors in the *Narcissus* extract. The MS spectrum revealed that the second peak is composed of at least two compounds, one of which was identified as galanthamine, a well-known AChE inhibitor. This example illustrates the great potential of bioresponse-linked instrumental analysis in the discovery of active compounds, particularly in combination with (electrospray) MS.

1.6   Conclusions

Configuring biochemical reactions on-line to liquid chromatography has proven to be highly advantageous over conventional microtiter plate based assays. The high sensitivity and selectivity of bioassays combined with the separation power of LC allows the rapid monitoring of reactive compounds in complex biological matrices. During the recent years the range of affinity interactions implemented in LC-BCD systems has expanded considerably, allowing the translation of a vast number of microtiter plate based assays into LC-BCD format. Consequently, bioanalysis and drug discovery, the two main application areas of on-line biochemical detection, can now be performed using a multitude of biologically interesting targets. In addition, the control of illegal hormones and environmental analysis represent other areas, which are highly suitable for LC-BCD. In conclusion, the methods described here allow automated monitoring of analytes, known or unknown, simultaneously with their biologically active degradation products at low concentration levels. In combination with mass spectrometry, bioactive compounds can be characterized in dependence of the response of the biochemical detection system.

Acknowledgements· The authors would like to thank Dr. J. Häggblad (Karo Bio, Huddinge, Sweden) and Dr. H. Bang (Erlangen University, Germany) for the generous supply of human estrogen receptor and monoclonal anti-leukotriene antibodies. Prof. K.-S. Boos (University of Munich, Germany) is acknowledged for the gift of Lichrosorb RP-ADS C18-alkyl diol columns. We are indebted to Dr. G.

32

Marko-Varga (Astra-Draco, Lund, Sweden) for supporting the development of on-line LC-BCD technologies. Gilson Medical Electronics (Villiers-le-Bel, France) is acknowledged for instrumental an software support. G. Häusl (Boehringer Mannheim, Penzberg, Germany) is acknowledged for technical support. LC Packings (Amsterdam, The Netherlands) is acknowledged for support on micro-LC equipment.

1.7   References

Arefyev, A.A., Vlasenko, S.B., Eremin, S.A., Osipov, A.P., Egorov, A.M. (1990): Flow-injection enzyme-immunoassay of haptens with enhanced chemiluminescence detection. Anal. Chim. Acta 237, 285-289.

Behrendt, M., Ploug, M., Ronne, E., Hoyer-Hansen, G., Dano, K. (1993): Cellular receptor for urokinase-type plasminogen activator: protein structure. Methods Enzymol. 223, 207-222.

Cassidy, S.A., Janis, L.J., Regnier, F.E. (1992): Kinetic chromatographic sequential addition immunoassays using protein A affinity chromatography. Anal. Chem. 64, 1973-1977.

Evans, M., Palmer, D.A., Miller, J.N., French, A. (1994): Flow injection fluorescence immunoassay for serum phenytoin using perfusion chromatography. M.T. Anal. Proc. 31(2), 7-8.

Freytag, J.W., Lau, H.P., Wadsley, J. (1984): Affinity-column-mediated immunoenzymometric assays: Influence of affinity-column ligand and valency of antibody-enzyme conjugates. J. Clin. Chem. 30, 1494-1498.

De Frutos, M., Regnier, M., Tandem, F. (1993): Chromatographic-immunological analysis. Anal. Chem. 65, 17A-25A.

Gelpi, E. (1985): Radioimmunoassay and the development of RIA-HPLC procedures: an updated literature survey. Trends Anal. Chem. 4, XII.

Gelpi, E., Ramis, I., Hotter, G., Bioque, G., Bulbena, O., Rosello, J. (1989): Modern high-performance liquid chromatographic-radioimmunoassay strategies for the study of eicosanoids in biological samples. J. Chromatogr. 492, 223-250.

Gosling, J.P. (1990): A decade of development in immunoassay methodology. Clin. Chem. 36, 1408-1427.

Gübitz, G. (1990): Separation of drug enantiomers by HPLC using chiral stationary phases-a selective review. Chromatographia 30, 555-564.

Gunaratna, P.C., Wilson, G.S. (1993): Noncompetitive flow injection immunoassay for a hapten, alpha-(difluoromethyl)ornithine. Anal. Chem. 65, 1152-1157.

Harajiri, S., Wood, G., Desiderio, D.M. (1992): Analysis of proenkephalin A, proopiomelanocortin and protachykinin neuropeptides in human lumbar cerebrospinal fluid by reversed-phase high-performance liquid chromatography, radioimmunoassay and enzymolysis. J. Chromatogr. 575, 213-222.

Heithier, H., Hallmann, D., Boege, F., Reilander, H., Dees, C., Jaeggi, K.A., Andt-Jovin, D., Jovin, T.M., Helmreich, E.J.M. (1994): Synthesis and properties of fluorescent beta-adrenoceptor ligands. Biochem. 33, 9126-9134.

Higazi, A., Cohen, R.L., Henkin, J., Kniss, D., Schwartz, B.S., Cines, D.B. (1995): Enhancement of the enzymatic activity of single-chain urokinase plasminogen activator by soluble urokinase receptor. J. Biol. Chem. 270, 17375-17380.

Ingkaninan, K., Hazekamp, A., de Best, C.M., Irth, H., Tjaden, U.R., van der Heijden, R., Verpoorte, R. (2000): HPLC with on-line coupled UV-MS biochemical detection for identification of acetylcholinesterase inhibitors from natural products. J. Chromatogr. A. 872, 61-73.

Irth, H., Oosterkamp, A.J., van der Welle, W., Tjaden, U.R., van der Greef, J. (1992): On-line immunochemical detection in liquid chromatography using fluorescent labeled antibodies. J. Chromatogr. 633, 65-72.

Krämer, P.M., Schmid, R.D. (1991a): Automated quasi-continuous immunoanalysis of pesticides with a flow-injection system. Pestic. Sci. 32, 451-462.

Krämer, P.M., Schmid, R.D. (1991b): Flow injection immunoanalysis (FIIA) - A new immunoassay format for the determination of pesticides in water. Biosens. Bioelectron. 6, 239-243.

Kuiper, J., Rijken, D.C., de Munk, G.A., van Berkel, T.J. (1992): In vivo and in vitro interaction of high and low molecular weight single-chain urokinase-type plasminogen activator with rat liver cells. J. Biol. Chem. 267, 1589-1595.

Kusterbeck, A.W., Wernhoff, G.A., Charles, P.T., Yeager, D.A., Bredehorst, R., Vogel, C.W., Ligler, F.S. (1990): A continuous-flow immunoassay for rapid and sensitive detection of small molecules. J. Biol. Methods 135, 191-197.

Lee, Y.J., Notides, A.C., Tsay, Y.-G., Kende, A.S. (1977): Coumestrol, NBD-norhexestrol, and dansyl-norhexestrol, fluorescent probes of estrogen-binding proteins. Biochem. 16, 2896-2901.

Liu, H., Yu, J.C., Bindra, D.S., Givens, R.S., Wilson, G.S. (1991): Flow injection solid-phase chemiluminescent immunoassay using a membrane-based reactor. Anal. Chem. 63, 666-669.

Locascio-Brown, L., Plant, A.L., Durst, R.A. (1988): Liposome-based flow injection immunoassay system. J. Res. Natl. Inst. Stand. Technol.(US) 93, 663-665.

Lutz, E.S.M., Irth, H., Tjaden, U.R., van der Greef, J. (1996): Applying hollow fibres for separating free and bound label in continuous-flow immunochemical detection. J. Chromatogr. A 755, 179-187.

Lutz, E.S.M., Irth, H., Tjaden, U.R., van der Greef, J. (1997): Implementation of affinity solid-phases in continuous-flow biochemical detection. J.

Chromatogr. A 776, 169-178.

Mattiasson, B., Nilsson, M., Berden, P., Hakanson, H. (1990): Flow-ELISA-binding assays for process-control. Trends Anal. Chem. 9, 317-321.

Miller, K.J., Herman, A.C. (1996): Affinity chromatography with immunochemical detection applied to the analysis of human methionyl granulocyte colony stimulating factor in serum. Anal. Chem. 68, 3077-3082.

Nelson, K., Pavlik, E.J., van Nagell, J.R., Hanson, M.B., Donaldson, E.S., Flanigan, R.C. (1984): Estrogenicity of coumestrol in the mouse: fluorescence detection of interaction with estrogen receptors. Biochem. 23, 2565-2572.

Newton, R.C., Uhl, J., Chang, J.Y., Lewis, A. (1989): Pharmalogical methods in control of inflammation. Alan R. Liss, New York.

Novokhatny, L., Medved, A., Mazar, P., Marcotte, J., Henkin, K., Ingham, K. (1992): Domain structure and interactions of recombinant urokinase-type plasminogen activator. J. Biol. Chem. 267, 3878-3885.

Oosterkamp, A.J., Beth, M., Unger, K.K., Irth, H., Tjaden, U.R., van der Greef, J. (1994a): Bioanalysis of digoxin and its metabolites using direct serum injection combined with liquid chromatography and on-line immunochemical detection. J. Chromatogr. B 653, 55-61.

Oosterkamp, A.J., Irth, H., Tjaden, U.R., van der Greef, J. (1994b): On-line coupling of liquid chromatography to biochemical assays based on fluorescent labeled ligands. Anal. Chem. 66, 4295-4301.

Oosterkamp.A.J., Irth, H., Heintz, L., Marko-Varga, G., Tjaden, U.R., van der Greef, J. (1996a): Simultaneous determination of cross-reactive leukotrienes in biological matrices using on-line liquid chromatography immunochemical detection. Anal. Chem. 68, 4101-4106.

Oosterkamp, A.J., Villaverde Herraiz, M.T., Irth, H., Tjaden, U.R., van der Greef, J. (1996b): Reversed-phase liquid chromatography coupled on-line to receptor affinity detection based on the human estrogen receptor. Anal. Chem. 68, 1201-1206.

Oosterkamp, A.J., van der Hoeven, R., Glasgen, W., Konig, B., Tjaden, U.R., van der Greef, J. (1998): Gradient reversed-phase liquid chromatography coupled on-line to receptor-affinity detection based on the urokinase receptor. J. Chromatogr. 715, 331-338.

Palmer, D.A., Ren, X.Z., Fernandez-Hernando, P., Miller, J.N. (1993): A model on-line flow injection fluorescence immunoassay using a protein A immunoreactor and lucifer yellow. Anal. Lett. 26, 2543-2553.

Ploug, M., Eriksen, J., Plesner, T., Hansen, N.E., Dano, K. (1992): A soluble form of the glycolipid-anchored receptor for urokinase-type plasminogen activator is secreted from peripheral blood leukocytes from patients with paroxysmal nocturnal hemoglobinuria. Eur. J. Biochem. 208, 397.

Schenk, T., Irth, H., Marko-Varga, G., Edholm, L-E, Tjaden, U.R., van der Greef, J. (submitted): On-line liquid chromatography-immunochemical detection of cytokines. Anal. Chem.

Shellum, C., Gübitz, G. (1989): Flow injection immunoassays with acridinium ester-based chemi-luminescence detection. Anal. Chim. Acta 227, 97-107.

Tang, H.T., Halsall, H.B., Heineman, W.R. (1991): Electrochemical enzyme immunoassay for phenytoin by flow-injection analysis incorporating a redox coupling agent. Clin. Chem. 37, 245-248.

Vetticaden, S.J, Barr, W.H., Beightol, L.A. (1986): Improved method for assaying digoxin in serum using high-performance liquidchromatography-radioimmunoassay. J. Chromatogr. 383, 187-193.

Whelan, J.P., Kusterbeck, A.W., Wernhoff, G.A., Bredehorsand, R., Ligler, F.S. (1993): Continuous-flow immunosensor for detection of explosives. Anal. Chim. Acta 65, 3561-3565.

Whicher, J.T., Evans, S.W. (1990): Cytokines in disease. Clin. Chem. 36, 1269-

1281.

Wittmann, C., Schmid, R.D. (1993): Application of an automated quasi-continuous immunoflow injection system to the analysis of pesticide-residues in environmental water samples. Sens. Actuators B15, 119-126.

Wortberg, M., Cammann, K., Strupat, K., Hillenkamp, F. (1994): Fresenius A new non-enzymatic tracer for a time resolved fluorescence immunoassay of triazine herbicides. J. Anal. Chem. 348, 240-245.

2 DETECTION OF BIOEFFECTIVE ENVIRONMENTAL COMPOUNDS BY HIGH-PERFORMANCE THIN LAYER CHROMATOGRAPHY

Christel Weins

Staatliches Institut für Gesundheit und Umwelt des Saarlandes, Malstatterstr. 17, 66117 Saarbrücken, Germany

Abstract. The effiency of a chromatographic analysis method depends on the selectivity of the chromatographic separation and the specifity of the detection method. In the case of High Performance Thin Layer Chromatography (HPTLC) the separated components can be detected and quantified directly on the chromatogram by physical and chemical methods. Coupling high performance thin-layer chromatography with biological or biochemical inhibition tests it is possible to detect toxicological active substances in situ

2.1 Pollutant analysis in the environment and the principle of "activity analysis"

The investigation of environmental samples, such as water, soil and air, for toxicologically relevant substances presents problems for every analytical technique.

The use of biological toxicity tests or enzymatic inhibition tests as screening procedures for the analysis of ground, drinking, surface and waste waters provides an initial indication of the presence of toxic pollutants in an environment. The results of this "biomonitoring" generally yield a summation of the damaging effects of pollutants in a defined test system; however, it is not possible to identify individual substances.

In order to demonstrate the presence of one or more pollutants, which are responsible for the toxic effect in the test system used above, it is necessary to resort to instrumental analysis, such as is carried out using gas chromatography or liquid chromatography (HPLC, HPTLC), with which it is now possible to detect the smallest traces of individual substance (e.g. in the ng to pg range).

For its part, the use of instrumental analysis requires previous selective enrichment of the active substance from the particular matrix: A fine analytical separation precedes identification with the aid of selected reference substances followed up by quantification of the active substance. Difficulties are encountered in the selection of relevant reference substances. It is only possible to detect those substances that are actively sought by the analyst. The analysis of individual substances does not detect unknown substances or metabolites, having adverse biological or toxicological effects, in an environmental sample.

2.2 Aims and fundamentals of activity analysis by thin-layer chromatography

The aim of activity analysis must be to detect and identify organic pollutants from environmental samples, having biological-toxicological activity in trace concentrations within the 100-200 ng/kg range. It is also necessary to detect and quantify a direct correlation with their toxic properties.

In order to be able to detect organic pollutants having biological-toxic effects, that are present in the environmental sample, it is necessary that the method selected should be as **universal** as possible. It is not the selectivity with respect to individual substances that is of importance for the method, but rather it is the detection of all or at least as many as possible of the organic pollutants, that are present in the environmental sample. This is the goal of the analyst.

Hence, the choice of sample preparation is of decisive importance for this procedure. The problem involved initially determines whether selective sample preparation shall be used to detect one active substance in the environmental sample or whether universal sample preparation shall be used to detect as many active substances as possible.

A further parameter, the enrichment factor, depends on the toxicity of the active substance. It is possible to detect very small traces of highly toxic substances by means of activity analysis, so that enrichment may not have to be carried out in these circumstances.

2.3 Principle of the method

Activity analysis involves a **coupling** together of two different methods. On the one hand, a pollutant analysis, using trace analytical methods, is used for the determination of selected organic pollutants and, on the other hand, the physical/chemical assessment is followed by a biological or biochemical toxicity test, thus, allowing a direct activity-dependent evaluation to be made after chemical/physical characterisation.

2.3.1 HPTLC as an instrument for pollutant analysis in the environment

Thin-layer chromatography (TLC) is one of the longest known and most thoroughly tested methods used in the analysis of environmental pollutants. As the analysis of plant protection agent residues began to become important in the 1950s the only methods available to the analyst were spectrophotometry and paper chromatography. However, these were soon replaced by thin-layer chromatography, which made it possible to separate numerous plant protection agents and to detect and identify them with various colour reagents (Walker and Beroza 1993).

Many publications appearing during the last years reveal clearly that the emphasis of environmental analysis is in the field of pollutant analysis such as pesticide analysis (Mori et al. 1994; Judge et al. 1993) with particular attention being paid to specific applications to insecticides, in particular to organophosphates (Patil and Shingare 1993), such as dichlorvos and its metabolites, whose concentrations have been determined in drinking water, in waste water, in beverages and in urine (Katagi et al. 1993).

Since the method of HPTLC allows a large sample throughput useful applications of thin-layer chromatography include the analysis of explosive substances in old military sites (Sokolowski and Rozylo 1993) and the investigations of aromatic amines (Ramachandran and Gupta 1993) and detergents (Kruse et al. 1994) in waste waters. Similar advantages apply to the determination of polycyclic aromatic hydrocarbons (PAH), which have been determined in soils (Baranowska et al. 1994; Furton et al. 1993), dust particles and aerosols (Tyrpien et al. 1994).

2.3.1.1   Instrumental analysis: Automated Multiple Development (AMD)

Since the plant protection law emphasises that drinking water has to be protected the drinking water regulation of 22.05.1986 set a limiting concentration of 0.1 µg/l for individual plant protection agents (PPA), the sum of all PPA must not exceed 0.5 µg/l; the permitted error is 50% of the limit This is a great challenge for any analytical method.

The determination of organic plant protection agents in drinking water with the AMD method according to DIN 38407, part 11 described a thin-layer chromatographic separation after enrichment of the contaminant by solid phase extraction.

This **universal method** involves the separation of the individual components by a stepwise multiple development on a normal stationary phase, whereby the elution commences with polar mobile phase and finishes with a nonpolar one. This allows the separation of substances of varying polarity, which is typical of the normal pollutants and their metabolites. Identification and determination is carried out by in situ reflectance measurement at 7 wavelengths (200 nm - 320 nm).

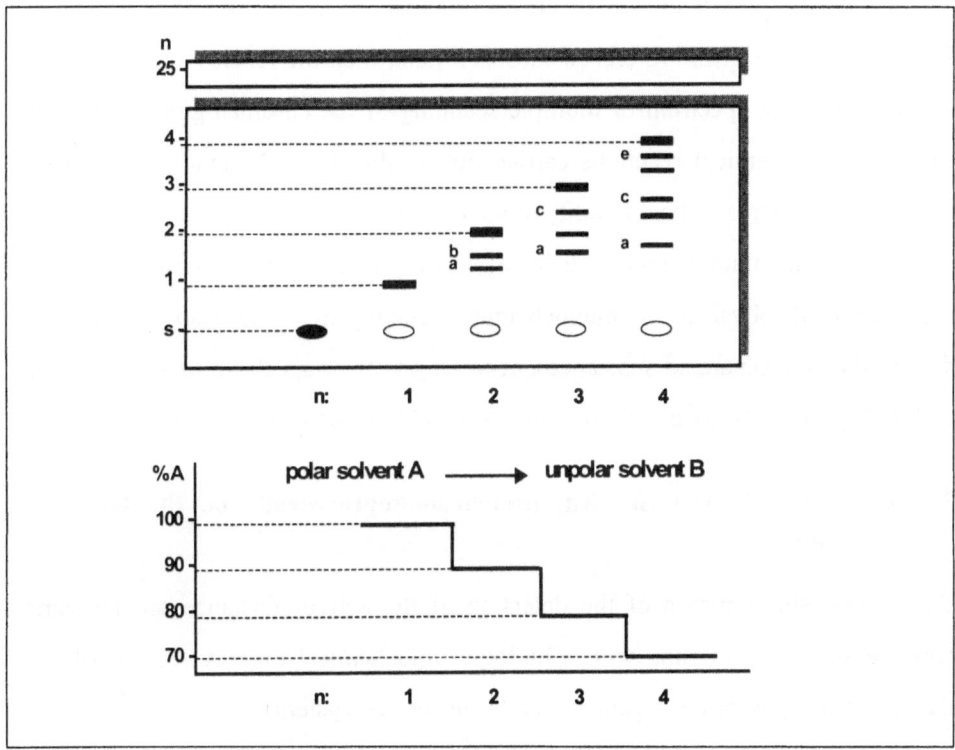

Fig. 2.1: Principle of the AMD gradient elution, N = number of step; a – e = components of the sample (Burger 1988)

The individual substances are subjected to preliminary identification – as in HPLC – on the basis of their position in the chromatogram and by comparison of the reflectance spectrum with a spectral library. Identification can then be secured by chromatographing the sample with a second gradient of different composition to confirm the result. Here the detection limit generally depends on the absorption coefficients of the substance or of its derivative, being tested.

By this method it is possible to separate, for example, 10 samples of unknown composition together with 10 test mixtures each containing 8 - 10 individual substances (i.e. a total of 80 - 100 reference substances) in parallel (Butz and Stan 1995).

Unknown pollutants are identified initially by comparing their retention factors with those of the reference substances and then the result is confirmed by means of the UV spectrum or multiple scanning of the chromatogram at various wavelengths. Chemical reactions carried out on the same chromatographic plate can be used to increase the reliability of identification.

Although unknown pollutants have generally been identified until now by the application of physical or microchemical techniques, it is also possible to differentiate the separated substances, according to their spectrum of action on the chromatogram itself, by the use of biological and biochemical test procedures.

2.3.1.2 Toxicity test (in situ) postchromatographically on the HPTLC-plate

The second step consists of the detection of the active substance on the same chromatogram by coupling with a biological/biochemical toxicity test, involving damage to an appropriate organism that forms the test system.

Here test organisms, such as mould spores, yeast cells, bacteria or cell organelles, such as chloroplasts, in a suitable nutrient medium are applied to the chromatogram. The biological signal, such as inhibition or stimulation of growth, inhibition or stimulation of luminescence or inhibition of photosynthesis, is used to localize toxic substances on the chromatogram or to detect unknown toxic substances. In addition to the use of organisms or sub-organisms as test materials it is also possible to use enzyme inhibition tests, as biochemical marker, for toxicologically relevant substances.

Documentation can be carried out by means of a flat bed scanner or a video camera and the inhibition can be reported quantitatively in defined toxicity units. The sensitivity here depends on the toxicity of the substance concerned on this defined test system.

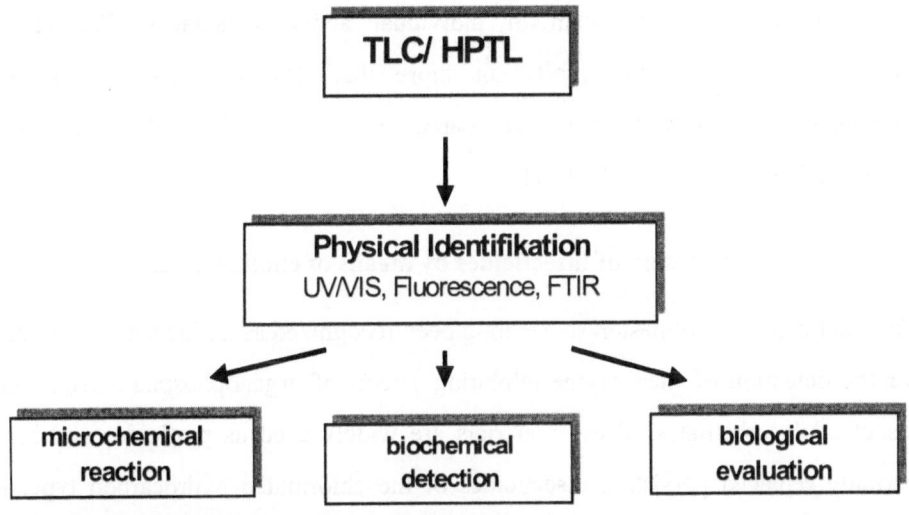

Fig. 2.2: Principle of "bioactivity based analysis" in HPTLC

2.4   Possible applications of activity analysis

Biological and biochemical test procedures are used to detect physically active pollutants in activity analysis. Specific enzyme inhibition tests on the thin-layer plate or test procedures involving organisms that use inhibition of bacterial growth, inhibition of bacterial luminescence or inhibition of the growth of a yeast strain as the signal serve to detect the presence of toxicologically relevant pollutants.

2.4.1 Enzyme-inhibition tests as detection methods in thin-layer chromatography

Mendoza and Shields (1973) published a summary of the application of enzyme-inhibition techniques in combination with thin-layer chromatography in 1973. They discussed the determination and identification of pollutants with insecticidal effects in residue analysis of vegetable, soil and water samples, together with the forensic toxicology of the metabolism of individual active substances. The report summarises the detection limits for more than 100 phosphate esters and carbamates. In 1981 Ambrus et al. compared six TLC methods for the detection of 188 pesticides (Ambrus et al. 1981).

2.4.1.1 The detection of insecticides by means of cholinesterase

The inhibition of cholinesterase has long been recognized as a biochemical method for the detection of the enzyme-inhibiting effects of organophosphate esters and insecticidal carbamates. These materials are widely used as pesticides and have partially replaced persistent insecticides of the chlorinated hydrocarbon type. In contrast to organochlorine compounds, phosphate esters and carbamates are frequently characterized by high toxicity, low stability towards hydrolysis and good biological degradation properties, together with relatively high water solubility and high mobility. The investigations were carried out using isolated enzymes and enzyme homogenates (Herzsprung 1991).

2.4.1.2   The physiological importance of cholinesterase

Acetylcholinesterase exerts a key function in the control of cholinergic stimulus transmission (c.f. chapter 5). It is localized at the synaptic cleft of the peripheral and central nervous system. There it ensures rapid hydrolytic degradation of the acetylcholine, that is produced during parasympathetic stimulation. Crystal structure analysis has recently made it possible to describe its function at the molecular level. The acetylcholinesterases are members of the family of serine hydrolases.

2.4.1.3   The molecular mechanism of cholinesterase inhibition

The inhibition of cholinesterase is the result of an irreversible phosphorylation or carbamylation of the serine OH groups in the active centre of the enzyme. The organophosphorus pesticides and carbamate insecticides inhibit the cholinesterase to very different degrees.

Most irreversible inhibitors inhibit the enzymatic reaction completely, frequently by formation of a covalent bond if their concentration exceeds that of the reacting groups in the enzyme.

Irreversible inhibition is expressed mathematically by the following equations, where k_i is the constant for the formation of the enzyme-inhibitor conjugate:

$$E + I \xrightarrow{\ k_i\ } EI$$

$$k_i = [\, L \cdot mol^{-1} \cdot min^{-1} \,]$$

The magnitude of the constants of inhibition k_i expresses the strength of the inhibiting effect of the inhibitor. Some pesticides and their metabolites exhibit differences in the ratios of their inhibition constants with respect to cholinesterase of up to 1:500.

Tab. 2.1: Detection limits of several organophosphates, carbamates and pentachlorphenol with cholinesterase and their inhibition constant

Active substance (enzyme inhibitor)	Detection limit	Inhibition constant $k_i = [L \cdot mol^{-1} \cdot min^{-1}]$
parathion (after oxidation)	0.045 ng	-
paraoxon-ethyl	0.013 ng	4.9×10^5
paraoxon-methyl	0.400 ng	2.2×10^4
mevinphos	0.200 ng	1.4×10^4
dichlorvos	0.200 ng	5.2×10^4
cabaryl	0.200 ng	2.7×10^4
aldicarb	0.400 ng	1.6×10^4
butoxycarboxim	0.100 ng	3.2×10^3
butocarboxim	0.800 ng	1.6×10^3
oxamyl	0.800 ng	1.4×10^5
pentachlorophenol	20.00 ng	1.0×10

Many organophosphorus derivatives, in particular thio- and dithiophosphate derivatives, only inhibit cholinesterase to a very slight degree. However, their inhibiting effects can be increased by a factor of up to 1000 by oxidizing them to the organophosphate analogues.

Some insecticides can be determined qualitatively and quantitatively by means of the cholinesterase inhibition test. Thio- and dithiophosphate esters are converted to their analogous, toxicologically active phosphonates, using Br_2 as the oxidizing agent. The inhibitory effect is determined by the reduction in the enzymatic hydrolysis of 1-naphthyl acetate to 1-naphthol and acetic acid followed by coupling of the 1-napthol to yield a violet-blue dye (diazonium salt, Echtblau B). White zones of inhibition on a coloured background are produced around toxicologically active substances (Geike 1969). The detection limit should be proportional to the inhibition constant of the particular substance and can lie in the lower picogram range (Weins and Jork 1996).

Other enzymes as well as cholinesterase can be used in toxicity tests for specific pollutants or classes on the thin-layer plate.

Tab. 2.2: Possible pollutant or pollutant classes and the specific enzyme system for detection

Enzyme system	Possible pollutant or pollutant class
chymotrypsin, trypsin, elastase, cholinesterase	insecticidal carbamates, insecticidal organophosphorus compounds, organochlorine compounds and their metabolites (Geike 1969)
urease, amylase, aminolaevulinic acid dehydratase	heavy metals and organometallic fungicides (Mendoza and Schields 1973)
vegetable peroxidase	quinones (Mendoza and Schields 1973)
catalase	2,4-dichlorophenol, hydroxylamine, monochloramine, nitrite (Mendoza and Schields 1973)

2.4.2   The detection of antibiotic activity in water samples

A significant increase has been observed in the numbers of antibiotic-resistant bacteria in the last decade. These can be simultaneously resistant to up to eight antibiotics. Antibiotic-resistant bacteria are more common in regions where antibiotics are in use. It has been demonstrated that large numbers of antibiotic-resistant bacteria are present in the environment. On the one hand, they are released directly into the environment during the application of slurry and dung from intensive animal rearing, and, on the other hand, they collect in water treatment plants arriving in the waste waters from clinical and domestic sources and from there they reach the environment in the treated waste water. However, it is not just antibiotic-resistant bacteria that are released, but also the antibiotics themselves. This has led to discussions of the question of how far antibiotics are involved in the selection of antibiotic-resistant bacteria, not only in hospitals and intensive rearing units but also after they have entered the environment and bring about an increase in numbers of antibiotic-resistant bacteria, in water treatment plants in particular (Feuerpfeil et al. 1999).

Measurement of the antibiotically active substances in waters acquires an increasing importance in this context.

After chromatography by the AMD technique *Bacillus subtilis* (BGA) is used as the indicator organism in the activity analysis that follows. The growth of the test organism on the thin-layer plate is inhibited by antibiotically active inhibitors and is indicated by the production of zones of inhibition. Detection is carried out by means of a bacterial vitality test, where the bacterial lawn on the thin-layer chromatogram is sprayed with an MTT tetrazolium salt (Hamburger and Cordell 1987; Dimenna et al. 1986).

The size of the zone of inhibition is determined by the amount applied, on the one hand, and by the specific activity of the substance, on the other hand.

2.4.3   Detection of fungicidal and herbicidal active substances in environmental samples

One of the most costly procedures forming part of the characterisation of water is the determination of individual substances, such as traces of plant treatment agents, by instrumental methods. Here it is only possible to detect those substances for whose presence it is tested. As described above, it is not possible, in the analysis of individual substances, to detect unknown substances or active metabolites with biological/toxicological activity. There are ca. 300 permitted active substances and the development of an analytical procedure for even one group of substances involves a great deal of time and expense, while the possibility still remains that the user will already have released an alternative product into the environment which is not detected by the measurement program that has been set up.

Here activity analysis with AMD/HPTLC acquires great importance because it constitutes an instrument capable of determining substances and unknown pollutants defined by the measurement program, e.g. that inhibit the growth of fungal spores or a yeast strain.

The yeast strain *Rhodotorula rubra* has proven itself a very suitable test organism, which colours the plate red when it is growing well. In the presence of a fungicidal active substance a white zone of inhibition is produced, whose size depends on the amount of fungicide applied and on its specific activity (Hostettmann et al. 1997).

Isolated chloroplasts or chloroplast fragments from plant juices can be applied to the thin-layer chromatogram in order to detect herbicidal active substances, which are introduced into the environment as inhibitors of photosynthesis. After chromatography, the active substances with herbicidal activity are revealed in the presence of dichlorophenolindophenol (DCPIP) as an electron acceptor and of light (Sackmauerová and Kovác 1978, Hermanns 1995).

2.4.4 Analysis of bioeffective environmental compounds by *Photobacterium phosphoreum* on the HPTLC-plate

The comparatively quick and low-cost bioassay with the luminescent marine bacterium *Photobacterium phosphoreum,* strain NRRL-B-11177, has gained considerable popularity for monotoring various industrial effluents and the toxicity

52

of different chemicals. Many toxic substances (nearly 1350 individual organic compounds) show an inhibition of the bioluminescence of *Photobacterium phosphoreum* and *Vibrio fisheri* in vitro (Kaiser and Palabricia 1991).

In HPTLC these substances have been identified postchromatographicly in situ on the plate by dipping the plate for 2 s into a suspension of bacteria and determination the difference of photone-emission using a cooled charged coupled device camera in a dark chamber (Weisemann et al. 1996, Eberz et al. 1996). A linear correlation between the inhibition of bioluminescence of *Photobacterium phosphoreum* and the concentration of an inhibitor could be shown. Under the condition described the detection limit of pentachlorphenol was found between 10 and 20 ng on the HPTLC plate while the detection limit of dichlorphenol could be observed in the range of 7.5 ng (Weins and Jork 1996).

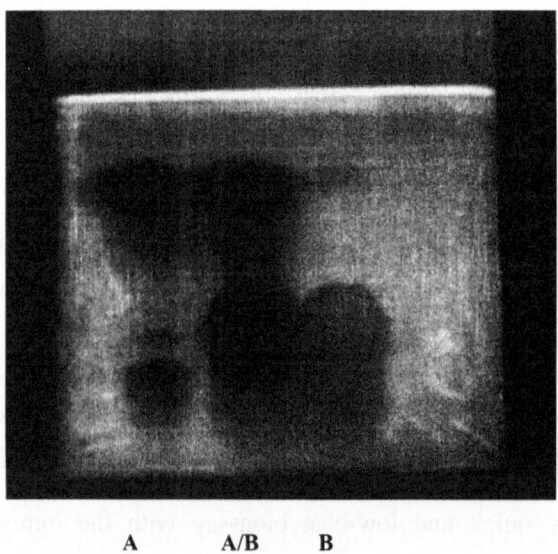

A   A/B   B

Fig. 2.3: Inhibition of the bioluminescence of *Photobacterium phosphoreum* postchromatographically in situ on the HPTLC-plate. A: Pentachlorphenol, 1 µg. B: 3´5´-Dichlorphenol, 1 µg

The biological detection in HPTLC can show that several working standards, such as pentachlorphenol or dichlorphenol are very often contaminated with toxic impurities, which can only be detected after chromatographic separation.

2.5   Value of activity analysis

Activity analysis is a coupling of instrumental analysis with biological/ biochemical activity tests. On the one hand, it is possible to assign physically detected substances to a selective activity, on the other hand, it is also possible to detect the presence of other unknown toxic active substances on the chromatogram. The universal gradient has been selected to optimize the chromatographic procedure so that it is possible to detect active substances of very different polarities, i.e. the actual substances and their active polar metabolites.

2.6   References

Ambrus, A., Hargitai, E., Caroly, G., Fulop, A., Lantos (1981): General method for determination of pesticide residues in samples of plant origin, soil and water. J. Assoc. Off. Anal. Chem. 64, 743-750.

Baranowska, I., Szeja, W., Wasilewski, P. (1994): Analysis of polycyclic aromatic hydrocarbons in soil extracts by adsorption and reversed-phase thin-layer chromatography. J. Planar Chromatogr. 7, 137-141.

Burger, K. (1988): Multimethode zur Ultraspurenbestimmung: Pflanzenschutz-mittel-irkstoffe in Grund- und Trinkwasser, analysiert durch DC/AMD (Automated Development). Pflanzenschutz-Nachrichten Bayer 41, 173-224.

54

Butz, S., Stan, H.-J. (1995): Screening of 265 pesticides in water by thin-layer chromatography with automated multiple development. Anal. Chem. 67, 620-630.

Dimenna, G.P., Walker, B.E., Turnbull, L.B., Wright, G.J. (1986): Thin layer bioautographic assay for salinomycin in chicken liver. J. Agr. Food Chem. 34, 472-474.

Eberz, G., Rast, H.-G., Burger, K., Kreiss, W., Weisemann, C. (1996): Bioactivity screening by chromatography-bioluminescence coupling. Chromatographia 43, 5-9.

Feuerpfeil, I., López-Pila, J., Schmidt, R., Schneider, E., Szewzyk, R. (1999): Antibiotikarestistente Bakterien und Antibiotika in der Umwelt. Bundesgesundheitsblatt-Gesundheitsforschung-Gesundheitsschutz 42, 37-50, Springer Verlag.

Furton, K.G., Jolly, E., Pentzke, G. (1993): Recent advances in the analysis of polycyclic aromatic hydrocarbons and fullerenes. J. Chromatogr. 642, 33-45.

Geike, F. (1969): Dünnschichtchromatographisch-enzymatischer Nachweis und zum Wirkungsmechanismus von Chlorkohlenwasserstoff-Insektiziden. J. Chromatogr. 44, 95-102.

Hamburger, O.M., Cordell, G.A. (1987): A direct bioautographic TLC assay for compounds possessing antibacterial activity. J. Natural Prod. 50, 25-29.

Hermanns, B. (1995): Wirkungsspezifische Detektion: Detektion von herbiziden Wirkstoffen in Dünnschichtchromatogrammen, Diplomarbeit an der Fachhochschule Aachen, Abt. Jülich (Prof. Dr. Jeromin).

Herzsprung, P. (1991): Methodische Grundlagen des Nachweises und der Bestimmung von insektiziden Phosphorsäureestern und Carbamaten im Wasser mittels Cholinesterasehemmung, Graduation Technische Universität München (Prof. Nießner).

Hostettmann, K., Terreaux, C., Marston, A., Potterat, O. (1997): The role of planar chromatography in the rapid screening and isolation of bioactive compounds from medicinal plants. J. Planar Chromatogr. 10, 251-257.

Judge, D.N., Mullins, D.E., Young, R.W. (1993): High-performance thin-layer chromatography of several pesticides and their major environmental by-products. J. Planar Chromatogr. 6, 300-306.

Kaiser, K.L.E., Palabricia, V.S. (1991): *Photobacterium phosphoreum*, Toxicity Data Index. Water Poll. Research J. Canada 26, 361.

Katagi, M., Tsuchihashi, H., Hanada, S., Himmori, H., Otsuki, K. (1993): Determination of dichlorvos, trichlorfon and their metabolites and degradation products in environmental water, drinks and human urine. Jpn. J. Toxicol. Environ. Health 39, 459-468.

Kruse, A., Buschmann, N., Cammann, K. (1994): Separation of different types of surfactant by thin-layer chromatography. J. Planar Chromatogr. 7, 22-24.

Mendoza, G.E., Schields, J.B. (1973): Determination of some carbamates by enzyme inhibition techniques using thin layer chromatography and colorimetry. J. Agric. Food Chem. 21, 178-184.

Mori, H., Sato, T., Nagase, H., Sakai, Y., Yamaguchi, S., Iwata, Y., Hashimoto, R., Yamazaki, F., Hayata (1994): Rapid screening method for pesticides as the cause substances of toxicosis by TLC. Jpn. J. Toxicol. Environ. Health 40, 101-110.

Patil, V.B., Shingare, M.S. (1993): Thin-layer chromatographic detection of organophosphorus insecticides containing a nitrophenyl group. J. AOAC-Int. 76, 1394-1395.

Ramachandran, K.N., Gupta, V.K. (1993): New analytical technique for the simultaneous determination of aromatic amines. Fresenius' J. Anal. Chem. 346, 457-458.

56

Sackmauerová, Kovác (1978): Thin layer chromatographic determination of triazine and urea herbicides in water by Hill-reaction inhibition detection technique. Fresenius Z. Anal. Chem. 292, 414-415.

Sokolowski, M., Rozylo, J.K. (1993): TLC analysis of warfare agents under battlefield conditions. J. Planar Chromatogr. 6, 467-471.

Tyrpien, K., Warzecha, L., Bodzek, D. (1994): Identification of PAH nitro derivatives in airborne particulate matter by TLC. Chem. Anal. (Warsaw) 39, 3.

Walker, K.C., Beroza, M. (1993): Thin-layer chromatography for insectizide analysis. J. Assoc. Off. Agric. Chem. 46, 250-261.

Weins, C., Jork, H. (1996): Toxicological evaluation of harmful stances by in situ biological and biochemical detection in high performance thin-layer chromatography. J. Chromatogr. 750, 403-407.

Weisemann, C., Kreiss, W., Rast, H.-G., Eberz, G. (1996): European Patent Application No. 0558139 A1.

3 ELECTRONIC NOSES

Wilhelm Boland, Dieter Spiteller

Max-Planck-Institut für Chemische Ökologie, Carl Zeiss-Promenade 10, 07745 Jena, Germany

Abstract. The vertebrate olfactory system is known for its extraordinary sensitivity and selectivity for odours. In recent years chemical sensors have been developed that are based on analogous distributed sensing properties. Instead of odorant receptor proteins instrumental olfaction utilises an array of solid-state sensors (chemoresistors, CHEMFETs, piezoelectric quartz microbalances). The sensor signals are processed by a pattern recognition engine (Artificial Neural Networks, ANN) to give a N-dimensional graphical readout representing the signature of an individual volatile or a given odorant blend (two dimensional VaporPrint^{TM}). High selectivity of sensors is attained by coatings of the detector surface with tailor-made host molecules. Macromolecular hosts, such as cage compounds, dendrimers, cyclodextrins, and libraries of cyclic peptides provide an almost unlimited molecular resource for problem-oriented selectivity-optimisation. In combination with gas chromatography electronic noses have become a rapid and powerful tool for environmental monitoring, smog control, general quality control, and investigative processes, such as drug- and explosives-discovery.

Did you ever try to measure a smell? Can you tell whether one smell is just twice as strong as another? Can you measure the difference between one kind of smell and another? It is very obvious that we have very many kinds of smells, all the way from the odour of violets and roses up to asafoetida. But until you can measure their likeliness and differences you have no science of odour. Alexander Graham Bell (1914).

3.1 Natural and instrumental olfaction

3.1.1 Functional comparison of human and instrumental olfaction

Olfaction often exhibits both high sensitivity for odours and high discrimination between them. To achieve this, the olfactory system makes use of feature detection using broadly tuned receptor cells organised in a convergent neurone pathway. Typically, the sensing mucosa of mammalians has some tens of millions of sensing cells which act as primary neurones and synapse with the secondary neurones in the olfactory bulb. On the average, 10-20,000 primary neurones synapse with a single secondary neurone resulting in a highly complex network (Figure 3.1a). Binding of the "odorants" creates signals in the primary neurones that are finally transmitted via the secondary neurones to the brain for further processing. It is the brain which interprets the signals and creates an impression what the sum of all these signals codes in terms of "odour". With respect to the general arrangement of processing units, artificial odour recognition by an "Electronic Nose" (Gardner 1987, Gardner and Bartlett 1999, Nagle et al. 1998)), as it is often called, works more like a human nose than conventional analytical instruments such as gas chromatographs or combinations of gaschromatographs with mass spectrometers (GC/MS) or infrared instruments (GC/FTIR). Unlike (gas) chromatographic techniques, the classical electronic nose doesn't attempt to separate or resolve a blend of volatiles into individual compounds. Analogous to the mammalian olfactory system it uses an array of different sensors that respond to each volatile in a slightly different way.

The output of the sensors (detectors) is then processed in a convergent feature detection system which has, at least in part, parallel feature detection. Like the mammalian organ, the artificial system does not require odour-specific receptors, but nevertheless it is able to process the ratio and the type of the signals to "identify" an odour; like the natural system it can be trained to "learn" to discriminate between smells.

a)

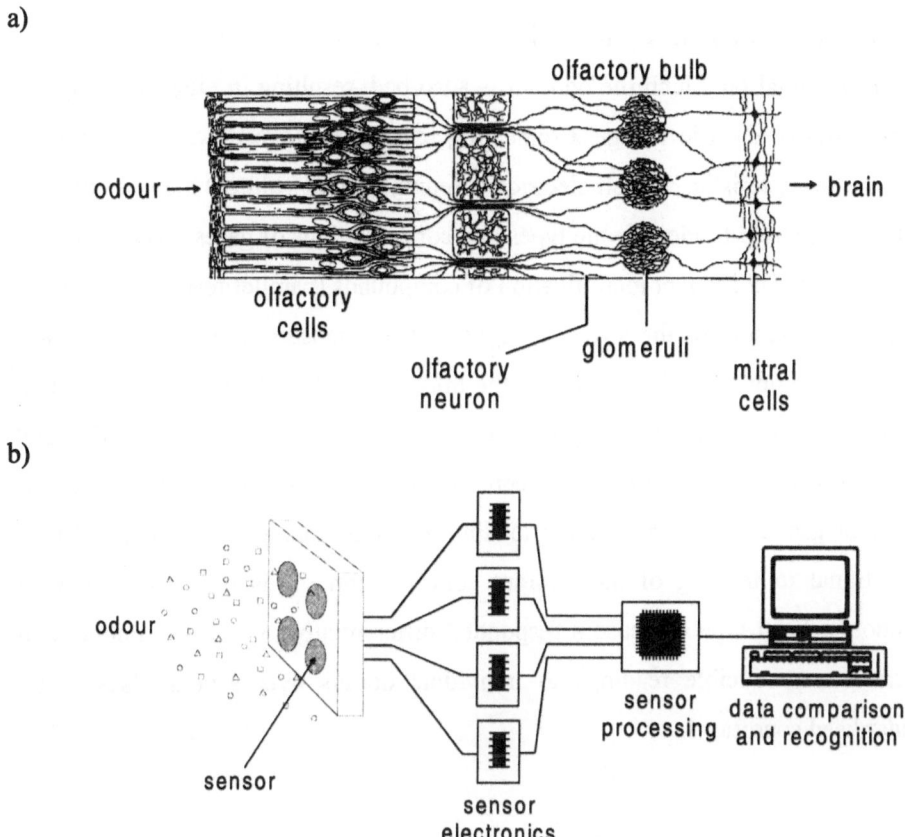

Fig. 3.1: Schematic comparison of human (a) and instrumental olfaction (b)

To construct an electronic model of a natural olfactory system (Figure 3.1a), an artificial device has to mimic the basic elements of the natural olfaction such as *i)* the odorant receptors responding to a wide range of chemical types (transducers) and *ii)* the convergent, parallel feature processing of the signals generated by the detectors (Figure 3.1b).

An electronic nose is, thus, an instrument, which comprises an array of electronic chemical sensors with partial selectivity and an appropriate pattern recognition system, capable of recognising simple or complex odours (Gardner et al. 1992).

In an electronic nose "recognition" is achieved via sensors incorporating a layer of material on which the odorant is adsorbed, resulting in either a change of conductivity or electric charge. Connecting several transducers, differing in their response characteristics into an electrical circuitry, output voltages may be obtained which can be used to identify the types and concentrations of gases and odorants in the gas phase. Subsequent identification of compounds (parallel feature processing) involves conversion of the analogue signals of the transducers into digital inputs for a microcomputer which processes the information. The first functional system based on such discrimination principles was described by Persaud and Dodd (1982). The primary sensors were represented by a simple array of three semi-conductor gas sensors, which, upon loading with odorants, gave a voltage change proportional to the log of the odorant concentration (as is the case in human olfaction!). Each type of sensors responded differentially to different odours and generated reproducible readings to individual odours over several days under standardised conditions.

3.1.2   Recognition of fragrance signatures

The individual sensors within an array ideally produce a time-dependent electrical signal in response to the quantity of a given odour component. Electronic noses with only a few sensors produce responses which are not correlated and multiple sensor elements eventually respond identical to the same vapour. For this reason uncorrelated sensor arrays can not identify single compounds but have to use artificial intelligence and Artificial Neural Networks (ANN) to, at least, recognise sensor patterns. ANNs are self learning; the more data presented, the more discriminating the instrument becomes. By running many standard samples and storing the results in computer memory, the application of ANN enables the electronic nose to "interpret" the significance of the sensor outputs.(Gardner et al. 1990). In this respect the electronic nose resembles the pattern recognition process of the human olfactory system. In fact, the ANNs processing elements (or nodes) can be compared to the neurones of the human brain (Figure 3.1a, b). "Learning" is achieved by varying the emphasis, or weight, that is placed on the output of one sensor versus another. The learning process is based on the mathematical distance between the obtained data sets, and, hence, large distances represent significant differences in sample-to-sample characteristics. The success of such cluster separation procedures is illustrated in Figure 3.2. Shurmer et al. (1989) applied this technique to diluted smoke from different brands of tobacco, mainly Burleigh and Virginian blends.

Figure 3.2a shows a scatter diagram formed by the raw digitised output of two sensors on a rectilinear grid for smoke from four brands of tobacco. The results for each brand overlap with those from at least one of the other brands. When each of the coordinates is divided by the standard deviation of the set to which it belongs and the results are replotted, one obtains the modified scatter diagram (Figure 3.2b), where each brand has become clearly identifiable.

Another powerful pattern recognition technique for the analysis of complex mixtures is the so called Multidimensional Scaling (Schiffman and Leffingwell 1981) which can transform objects into a two-dimensional spatial configuration; the procedure is well suited for the analysis of aromas and taste images.

3.2   Sensors

Since the beginning in the 1970s a multitude of sensors to detect specific gases and vapours have been developed. Table 3.1 compiles some common types of sensors

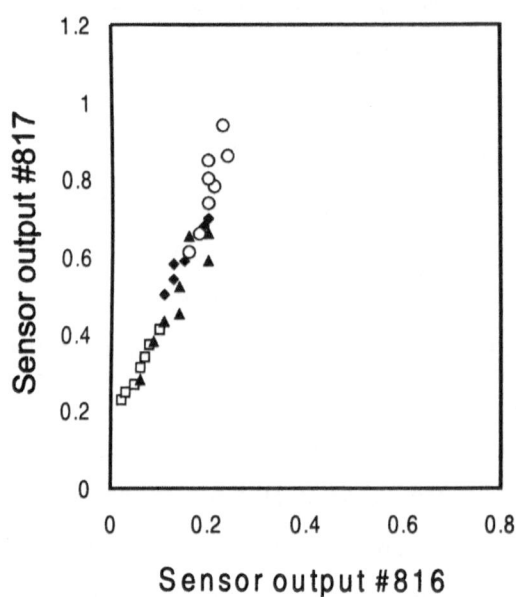

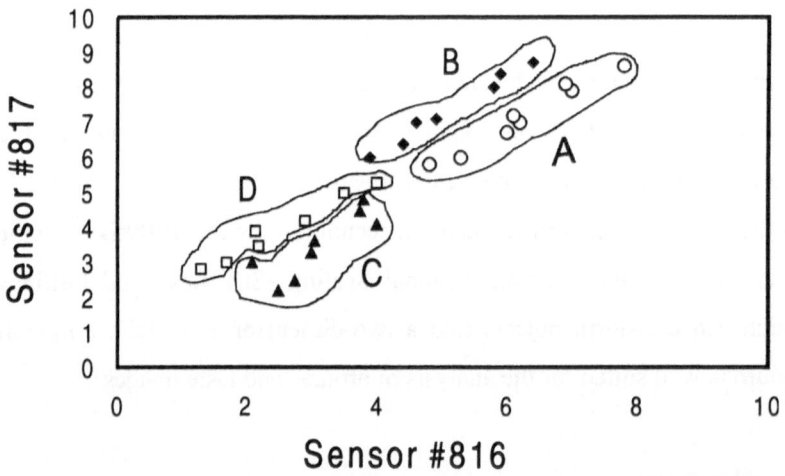

Fig. 3.2: Scatter diagram for four brands of tobacco. (a) Raw data, (b) Processed data

employed in multisensor arrays of electronic noses (cf. Figure 3.1b). Common to all is a broad response to a range or class of gases rather than a single compound.

Chemoresistors: Metal oxide sensors demonstrate good sensitivity to organic vapours (ppm or even ppb detection limits) for a very broad range of chemical compounds. They are quite sensitive to combustible materials such as alcohols, but are generally poor at detecting sulfur- or nitrogen-based compounds. The sensors are made by depositing a thin layer of oxide film on a ceramic material. By adding small amounts of other metal ions (Cr, Mn, Pd, etc.) during fabrication, sensors with different characteristics become available (Weimar et al. 1990). For proper functioning the sensors are usually heated between 100° and 600 °C. Their electrical resistance decreases in the presence of an odour, with the magnitude of the response dependent on the nature of the detected molecule and the type of the metal oxide used. The odorants react with chemisorbed oxygen species and, thus modulate the conductivity. Response time of metal oxide sensors is between 10 and 120 seconds.

Table 3.1: Sensors for detecting gases and organics

Active material	Sensor type	Target gases
Sintered metal oxide	Chemoresistor	Combustible gases
Catalytic metal	Thermal, e.g. pellistor, CHEMFETs	Combustible gases
Lipid layers	Acoustic, e.g. piezoelectric	Organics
Phthalocyanins	Chemoresistor, CHEMFET	NO_x, H_2, NH_3, HCl, SO_2, H_2S
Calixarenes, dendrimers	Quartz microbalance (QMB)	Organics
Conducting polymers	Chemoresistor, QMB, CHEMFETs, SGFETs	NH_3, alcohols
Electrochemical	Potentiometric/amperometric	NH_3, CO, ethanol
Organic semiconductors	Optical (e.g. IR)	CH_4, CO_2, NO_x

CHEMFETs: In this type of detectors the interaction of the sensing layer with an analyte modulates the workfunction of the layer, finally resulting in a shift of the transistor threshold voltage in a concentration dependend manner. At high gas concentrations the sensor response becomes logarithmic. Therefore, these sensors have a particular wide dynamic range and are well-suited for monitoring gas concentration fluctuating over a large range (Bartlett et al. 1989). The sensing mechanism of SGFETs with Pd- or Pt-based hydrogen sensors relies on the catalytic decomposition of hydrogen-containing molecules (H_2, HN_3, HCN, Cassidy et al. 1986). By doting conducting polymers (Polyanilin (PANI), PPy) with Pd-clusters or Hg, their sensing properties can be extended to otherwise unreactive gases such as H_2 or HCN.

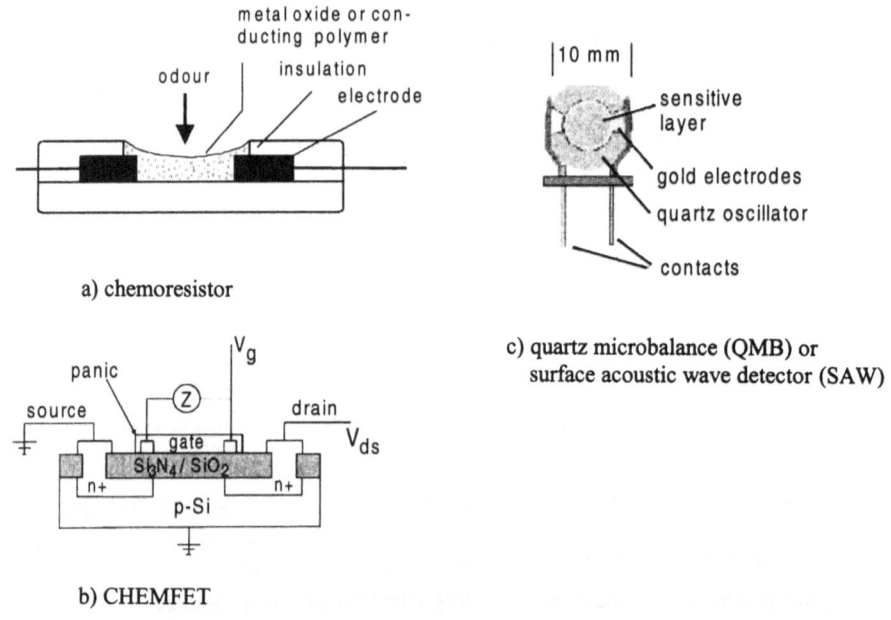

a) chemoresistor

b) CHEMFET

c) quartz microbalance (QMB) or surface acoustic wave detector (SAW)

Fig. 3.3: Principal construction of chemoresistors, CHEMFETs and quartz microbalances

Surface Acoustic Wave Crystals (SAWs): The most versatile electronic sensor is the piezo quartz microbalance (QMB; or surface acoustic wave (SAW) crystals). The QMB consists of a quartz crystal oscillating in the thickness shear mode at its fundamental frequency (Figure 3.3c). Any change in mass on the surface of the crystal results in a frequency shift that is easily recognised by appropriate electronic devices (Carey and Kowalski 1986). A common approach is to expose an array of SAW crystals with different polymer coatings to the vapour to be characterised. In theory, each polymer will absorb the vapours differently and, hence, each coated SAW will create a characteristic response curve. Because of the need for the analyte to diffuse into and out of the coating, coated crystals suffer from rather long analysis time. Uncoated SAW are very fast and highly sensitive (picogram sensitivity) and are, therefore often employed as detectors for electronic noses coupled to rapid chromatographic techniques.

Other methods of detection comprise electrochemical, spin-coating (polymers), screen-printing (phthalocyanins, Chadwick et al. 1986) and Langmuir-Blodgett (LB-films, Ohnishi et al. 1992) approaches (lipid coatings, phthalocyanins). It was found that films consisting of about 10 layers were sensitive to a variety of substances such as methanol, ethanol, methylamine, acetone, ammonia and tobacco smoke (Imanpour 1986). In recent years crystalline microporous Sn(IV) chalgogenides exhibiting zeolithe-type architecture have been obtained by organic template mediated hydrothermal synthesis. A remarkable property of these so called R-SnX-1 materials is the ability of their open frameworks to undergo elastic deformations in response to the inclusion of adsorbed molecular guests such as H_2O, H_2S or H_2Se (Ahari et al. 1995). Integration of these materials into transistors or diode array-type of chemosensors makes these materials highly attractive for electronic noses (Ozin 1995).

3.2.1 Organic sensor materials

Organic materials (lipid layers, conducting polymers etc.) widen the spectrum of sensors significantly. First, there is a much wider choice of materials and, second, appropriate functional groups that interact with different classes of odorant molecules can be selected and build into the active material. Conducting-polymer based sensors (Chemoresistors and CHEMFETs) can detect odorants with a molecular weight range of between 30 and 300 – essentially the same range that the human nose can detect. The resins work best with polar or charged compounds. Particularly widespread are electrochemically grown polymers made with polypyrrole (PPy, O'Riordan and Wallace 1986, Bartlett et al. 1989), polyaniline (PANI) and polythiophene (PT) resins (Waltman et al. 1983). By changing the solvent system or the counter ion associated with their manufacture, it is possible to modify the polymer chain and get subtle differences in the reactivity of each sensor. Upon absorption of volatile components onto the surface of the conductive polymers, they change their base resistance in a characteristic, substance-dependent fashion. The sensors can be operated close to room temperature (20°-60 °C) and have a sensitivity of typically 0.1-100 ppm. Conductive polymers can be also used for the fabrication of potentiometric sensors. Examples are a odorant sensitive field-effect transistors (CHEMFETs, Figure 3.3b) with a polyaniline gate (PANIFET, Liess et al. 1996) and suspended-gate (Pt, Pd and Pd-alloys) field-effect transistors (SGFET, Zhang et al. 1993). Metallo-porphyrins have been introduced as coatings on piezoelectric quartz microbalance (QMB) to obtain chemical sensors. The main advantage of such porphyrin-coated QMB-sensors is the dependence of the sensing properties on the nature of the central metal and on the lateral groups; both can be easily modified by synthesis to obtain optimal sensitivity.

3.2.2   Supramolecular host-guest systems

Supramolecular host-guest systems promise to be the most versatile materials for future developments in sensor technology. Lactam macrocycles (Behm et al. 1985, Vögtle and Hoss 1992), cyclodextrins (Dickert et al. 1995), (homo)calixarenes (Ibach et al. 1999) and dendrimers (Jansen et al. 1994) can be easily coated onto QMBs and used for selective recognition of analytes.

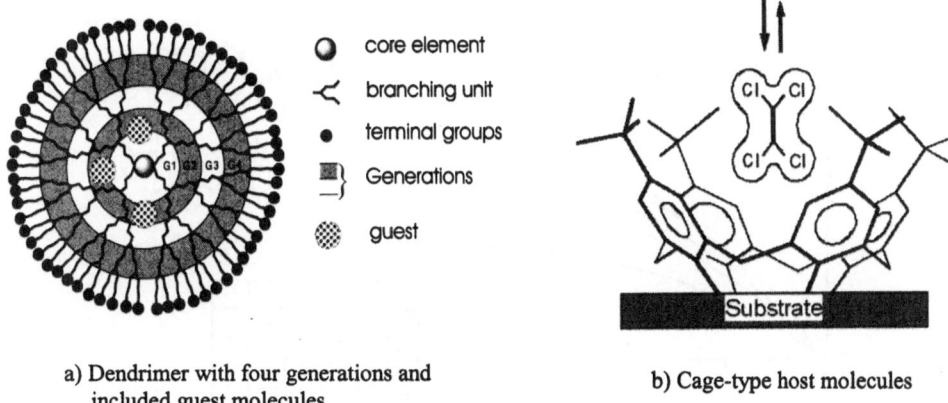

a) Dendrimer with four generations and included guest molecules	b) Cage-type host molecules

Fig. 3.4: Supramolecular hosts

In general, many different sensor-elements are integrated on a single quartz substrate, but each of them is individually connected to an resonating oscillator system allowing direct measurement of the mass change implied in the docking process. Typical interactions between guests and hosts are based on weak *van der Waals* forces, charge-transfer and a host/guest characteristic set of hydrogen bonding. Due to these weak interactions, the inclusion of guests into the macromolecular host is reversible and, hence, the corresponding sensors are useful for continuous monitoring. In case of the dendrimers, the efficiency of the incorporation of guests depends on the number of generations, since the space for analyte molecules increases with the host's size. Owing to the unspecific interactions of hosts and guests in conjunction with the simple, iterative build-up of the macromolecules, the hosts can be fine-tuned concerning cavity space, number of hydrogen bonding groups, electronic character to meet specific requirements, such as the distinction between aldehydes and ketones or substitution pattern of aromatics (Heil et al. 1999). An example for an efficient synthetic tuning of a host, exemplified with a lactam macrocycle by minimal change of structural elements (substitution of O by S; $X_1 \rightarrow X_4$), is illustrated in Figure 3.5.

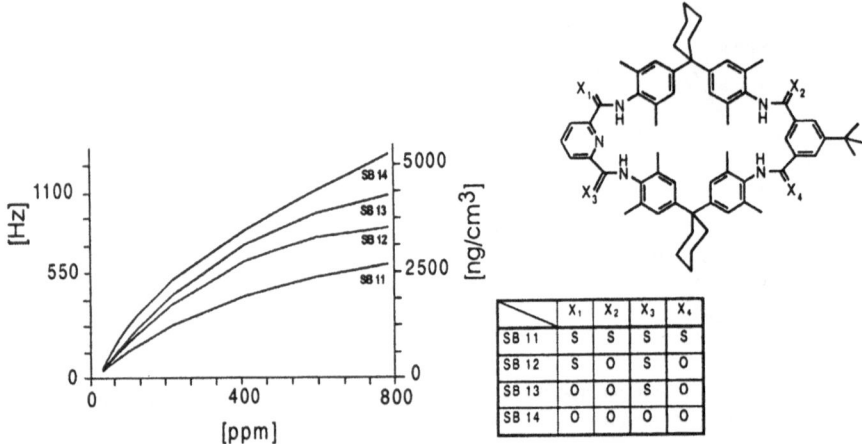

Fig. 3.5: Characteristic curves and molecular architecture of macrolactam-based sensor materials systematic exchange of amide into thioamides generates individually responding sensor materials (SB11 → SB14). Shown are their characteristic responses towards benzaldehyde.

3.2.3   Peptide libraries

Compared to the still limited number of cyclodextrins and calixarens, cyclopeptide libraries, available from commercial educts by combinatorial chemistry, allow the rapid build up of a very large population (about 10^{16}) of different molecules. Cyclopetides can be synthesised in different ring sizes, backbones, and side chain modifications. They can be designed to exhibit different molecular properties such as conformational restrictions, lipophilic or hydrophilic cavities. The compounds are easily coated onto QMBs and label attachment is generally without problems. Following the same general procedure, mirror images of the synthetic receptors can be made; an often indispensable feature for the reliable recognition of enantiomers. In combination with pattern recognition, cyclopeptide collections seem to be ideal biomimetic receptors which may be used in form of microstructured peptide-functionalised surfaces. Peptide-coated areas may also serve as matrix-supported planar bilayers for reconstituted biological receptors as sensor elements (Göpel and Heiduschka 1994, Göpel 1996).

3.2.4 Optical sensors

More recent sensor developments employ arrays of optical sensors to detect odo-rants (Dickinson et al. 1996). As shown in Figure 3.6 the sensor-arrays consists of a bundle of fibres coated at the distal tip with different polymer/dye deposits em-bedded in a permeable polymer membrane. Exposure to vapour induces changes in fluorescence intensity at a given wavelength which are then recorded and plotted versus time to produce a temporal response pattern for each sensing site to a given odour.

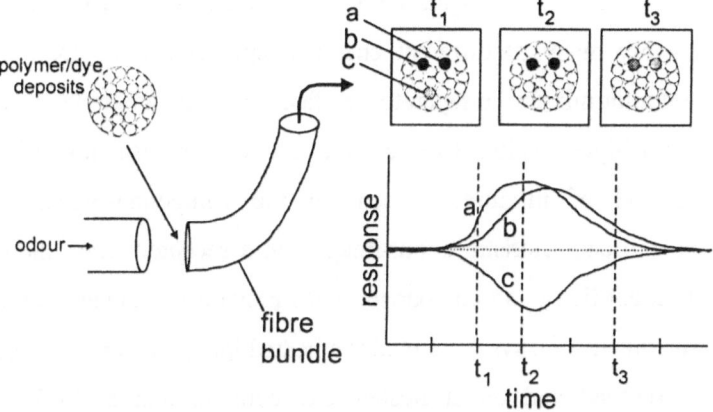

Fig. 3.6: Fibre glass array of optical sensors for odour detection. Pulses of air/solvent dilutions are delivered directly to the distal ends of the fibres in the sensor bundle. Fluorescence is plotted versus time to produce a series of patterns that are introduced into the neural pro-cessing system

Responses of the sensors remain stable over long periods and over many hundreds of analyte applications. The response diversity enables the system to discriminate between different odours using a limited number of sensing elements. To improve the rather low sensitivity (100-1000 ppm), common adsorbents such as alumina can be incorporated into the sensing matrix, lowering the vapour detection limit by nearly two orders of magnitude. The enhanced performance of these sensors is most likely due to both the physisorptive traits of alumina and the large surface area of the porous particles. Coated sensor bundles have been used successfully to discriminate between large numbers of analytes, from pair of enantiomers to different types of fragrances (Johnson et al. 1997).

3.3   Fast gas chromatography as an electronic nose

The classical electronic nose simultaneous interacts with all components of a blend. No attempt is made to separate complex mixtures into individual compounds. Thus, an electronic nose provides a recognisable visual image of specific vapour mixtures containing possibly many of different chemical species (cf. Figure 3.2a/b). If, however, the odorant mixture is separated by fast gas chromatography prior to exposure to a non-analyte specific SAW crystal, the detector responds to individual compounds instead of mixtures. Based on the compound-typical retention parameters and the characteristic odour-images on a receptor array, the odorants can be reliably identified upon comparison with previously trained data bases. In commercial instruments the system includes a heated inlet, vapour preconcentrator, a temperature ramped and direct heated GC column, and a SAW detector. Sensitivity is excellent because the uncoated SAW detector has pg sensitivity and there is no dilation of vapour sample. The system inlet can sample ambient air, desorbed vapour samples, or headspace vapours from liquid samples. The SAW crystal produces a variable frequency in response to analytes condensing and evaporating on the surface. In fast GC the chromatogram duration is 10 s, peak widths are in msec, and retention time is resolved to within 20 ms. Thus, up to 500 sensors in 10 s can be polled serially (Staples 1999). The sensor responses are nearly orthogonal with minimum overlap. A polar plot of chromatogram time with the radial direction being the sensor signal or the derivative of the sensor signal provides a graphical feature well suited for electronic noses pattern recognition algorithms (VaporPrintTM images). Analysis systems, based on SAW-sensors and miniature chromatographic columns have been used to detect drugs, explosives, volatile organics, polychlorinated biphenyls, and dioxins.

3.4   Applications

3.4.1   Fruit flavour analysis

One of the most important objectives of food industry is that of achieving a uniform quality of the final product such as the systematic determination of fruit ripeness under harvest and post-harvest conditions. Typical markers of the degree of ripeness are skin colour, sugar content, pH, ethylene and development of flavour. Due to their non-destructive analysis mode, electronic noses are particularly well suited for determining fruit ripeness. Equipped with sensors responding to ethylene (the ripening hormone in climacteric fruits, such as apples, peaches, bananas, etc.) and other volatiles emitted by fruits during ripening, the electronic nose can be used to follow the ripening process over the season. As the sensor signals are processed by a pattern recognition engine that allows the system to analyse complex aromas, their change in time and concentration can be easily followed and attributed to different categories of ripening. A typical instrumental set-up for on-line analyses from ripening fruits is shown in Figure 3.7.

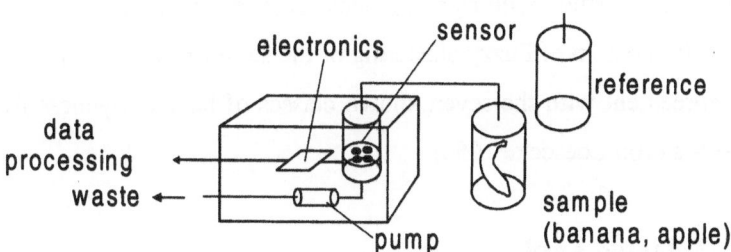

Fig. 3.7: Analysis of flavour blends from ripening fruits

The headspace from the enclosed fruits and that from an empty reference vessels were alternatively swept by a pump into the electronic nose (four tin-oxide sensors) and monitored (Llobet et al. 1999, Hines et al. 1999). The conductance was found to increase as the fruits ripened. Subsequent processing of the raw-data with neural networks (e.g. fuzzy ARTMAP, Carpenter et al. 1995, Hines et al. 1999) resulted in a clear and accurate classification of ripeness which correlated well with the peel colour as illustrated in Figure 3.8.

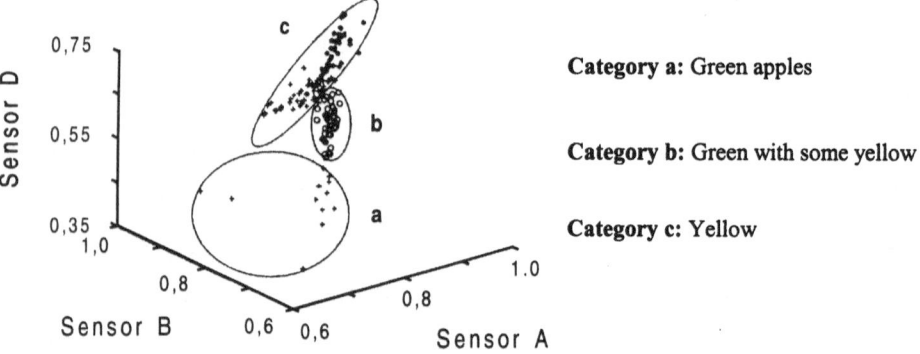

Fig. 3.8: Clusters representing three different categories of state of ripeness of apples

A comparable study with ripening bananas revealed seven different ripening categories with the help of Fuzzy clustering of the sensor responses. This finding is in perfect agreement with the seven known classes of banana ripeness from skin-colour analysis (von Loesecke 1950).

3.4.2   Selected applications

There are many important environmentally and safety related applications for an electronic nose. Hazardous gases, air pollution, and many types of odour are common pollutants or markers which require monitoring and quantification. The odour images (VaporPrint^{TM}) obtained by electronic nose techniques from flammable fuels, drugs of abuse, explosives, pollutants, ripening fruits, or even infectious bacteria in ground meat, are very rapidly available at moderate instrumental costs. These images together with the ability to detect an almost unlimited number of volatile chemicals leads to the wide diversity of applications. Applications are whereever there are vapours, odours, smells or fragrances to be measured. Table 3.2 gives a brief synopsis of published applications.

Table 3.2: Selected applications of electronic nose techniques

Test compound	Application	Sensor	Reference
Trimethylamine	Fish freshness	Metal oxides	Egashira et al. 1990
Volatiles	Apple ripeness	Metal oxide	Hines et al. 1999
Gases (CO,CH$_4$, H$_2$)	Monitoring of flammable gases	Metal doted SnO$_2$	Weimar et al. 1990
Gases (H$_2$, NH$_3$, air)	Monitoring of flammable gases	CHEMFETs	Domanský et al. 1998
Gases (HCN)	Monitoring of toxic gases	SGFETs	Li et al. 1993
N-Methylpyrrolidone	Toxic solvent	QMB	Malley et al. 1999
Volatiles	Food production	QMb	Nitz et al. 1999
Volatiles	Waste water	Ppy-conducting polymer	Fenner et al. 1999
Volatiles	Perfumery		Horner 1993
Volatiles	Perfumery	QMB	Nakamoto et al. 1993
Volatiles	Coffee	Metal oxide	Gardner et al. 1992
Volatiles	Beer	Conducting polymer	Pearce et al. 1993
Volatiles	Ground meat	CHEMFETs and SnO$_2$	Winquist et al 1993
Solvents	Isomeric xylenes	Calixarens, QMB	Dickert et al. 1999
Dioxines, furanes	Environmental monitoring	GC-SAW	Staples 1998

74

3.5   References

Ahari, H., Bedard, R.L., Bowes, C.L., Jiang, T., Lough, A.J., Ozin, G.A., Petrov, S., Young, D. (1995): Nanoporous Tin(IV) chalcogenides: Flexible open-framework nanomaterials for molecular discrimination. Adv. Mater. 7, 375-378.

Bartlett, P.N., Archer, P.B.M., Ling-Chung, S.K. (1989): Conducting polymer gas sensors. I: fabrication and characterization. Sens. Actuators 19, 125-140.

Behm, F., Simon, W., Müller, W.M., Vögtle, F. (1985): Macrocyclic oligolactams based on tetraphthalic acid as ionophores with selectivity depending on included guest molecules. Helv. Chim. Acta 68, 940-944.

Carey, W.P., Kowalski, B.R. (1986): Chemical piezoelectric sensor and sensor array characterization. Anal. Chem. 58, 3077-3084.

Carpenter, G., Grossberg, S., Reynolds, J. (1995): A fuzzy ARTMAP nonparametric probability estimator for nonstationary pattern recognition problems. IEEE Trans. Neural Networks 6, 1330-1336.

Cassidy, J., Pons, S., Janata, J. (1986): Hydrogen response of palladium coated suspended gate field-effect transistor. J. Anal. Chem. 58, 1757-1761.

Chadwick, A.V., Dunning, P.B.M., Wright, J.D. (1986): Application of organic solids to chemical sensing. Mol. Cryst. Liq. Cryst. 134, 137-153.

Dickert, F.L., Geiger, U., Keppler, M., Reif, M., Bulst, W.E., Knauer, U., Fischerauer, G. (1995): Supramolecular receptors for SAW and QMB devices – solvent recognition evaluated by FT-IR spectroscopy and BET adsorption analysis. Sens. Actuators B 26, 199-202.

Dickert, F.L., Hayden, O., Zenkel, M.E. (1999): Detection of volatile compounds with mass-sensitive sensor arrays in the presence of variable ambient humidity. Anal. Chem. 71, 1338-1341.

Dickinson, T.A., White, J., Kauer, J.S., Walt, D.R. (1996) : A chemical-detecting system based on a cross-reactive optical sensor array. Nature 382, 697-700.

Domanský, K., Baldwin, D.L., Grate, J.W., Hall, T.B., Li, J., Joscowicz, M., Janata, J. (1998): Development and calibration of Field-Effect Transistor-based sensor array for measurement of hydrogen and ammonia gas mixtures in humid air. Anal. Chem. 70, 473-481.

Egashira, M., Shimizu, Y., Takao, Y. (1990): Trimethylamine sensor based on semiconductive metal oxides for detection of fish freshness. Sens. Actuators B1, 108-112.

Fenner, R.A., Stuetz, R.M. (1999): The application of electronic nose technology to environmental monitoring of water and wastewater treatment activities. Water Environ. Res. 71, 282-289.

Gardner, J.W. (1987): Pattern recognition in the Warwick Electronic Nose, 8th Int. Congress of European Chemoreception Research Organisation, University of Warwick, UKL, July 1987.

Gardner, J.W., Hines, E.L., Wilkinson, M. (1990): Application of artificial neural networks in an electronic nose. Meas. Sci. Technol. 1, 446-451.

Gardner, J.W., Shurmer, H.V., Tan, T.T. (1992): Application of an electronic nose to the discrimination of coffees. Sens. Actuators B6, 71-75.

Gardner, J.W., Bartlett, P.N. (1999): Electronic noses: principles and applications. Oxford University Press.

Göpel, W., Heiduschka, P. (1994): Introduction into bioelectronics: interfacing biology with electronics. Biosens. Bioelectronics 9, 3-13.

Göpel, W. (1996): Ultimate limits in the miniaturization of chemical sensors. Sens. Actuators. 56, 83-102.

Heil, C., Windscheif, G.R., Braschoss, S., Flörke, J., Gläser, J., Lopez, M., Müller-Albrecht, J., Schramm, U., Bargon J., Vögtle, F. (1999): Highly selective sensor materials for discriminating carbonyl compounds in the gas phase using quartz microbalances. Sens. Actuators B61, 51-58.

Hines, E.L., Llobet, E., Gardner, J.W. (1999): Neural network based electronic nose for apple ripeness determination. Electronic Lett. 35, 821-823.

Horner, G. (1993): in: Sensor 93, Kongress 11.-14. Oktober Messezentrum Nürnberg Volume 2 „Einsatz von Sensor-Arrays in der Olfaktometrie", 179.

Ibach, S., Prautsch, V., Vögtle, F., Chartroux, C., Gloe, K. (1999): Homocalixarenes. Acc. Chem. Res. 32, 729-740.

Imanpour, M. (1986): Development of processing conditions for production of Langmuir-Blodgett films for an electronic nose. MSc disseration, University of Warwick.

Jansen, J.F.G.A., de Brabander-van den Berg, E.M.M., Meijer E.W. (1994): Encapsulation of guest molecules into a dendritic box. Science, 226, 1226-1229.

Johnson, S.R., Sutter, J.M., Engelhardt, H.L., Jurs, P.C., White, J., Kauer, J.S., Dickinson, T.A., Walt, D.R. (1997): Identification of multiple analytes using an optical sensor array and pattern-recognition neural networks. Anal. Chem. 69, 4641-4648.

Li, J., Petelenz, D., Janata, J. (1993): Suspended gate field-effect transistor sensitive to gaseous-hydrogen cyanide. Electroanal. 5, 791-794.

Liess, M., Chinn, D., Petelenz, D., Janata, J. (1996): Properties of insulated gate field-effect transistors with a polyaniline gate electrode. Thin Solid Films 286, 252-255.

Llobet, E., Hines, E.L., Gardner, J.W., Franco, S. (1999): Non-destructive banana ripeness determination using a neural network-based electronic nose. Meas. Sci. Technol. 10, 538-548.

Malley, L.A., Kennedy, G.L., Elliott, G.S., Slone, T.W., Mellert, W., Deckhardt, K., Gembardt, C., Hildebrandt, B., Parot, R.J., McCarthy, T.J., Griffith, J.C. (1999): 90-day subchronic toxicity study in rats and mice fed N-methylpyrrolidone (NMP) including neurotoxicity evaluation in rats. Drug Chem. Toxicol. 22, 455-480.

Nagle, H.T., Schiffman, S., Guiterrez-Osuna, R. (1998): The how and why of Electronic noses. IEEE Spectrum 35, 22-23.

Nakamoto, T., Fukuda, A., Moriizumi, T. (1993): Perfume and flavour identification by odour sensing system using quartz-resonator sensor array and neural-network pattern-recognition. Sens. Actuators B10, 85-90.

Nitz, S., Kollmannsberger, C., Lachermeier, G., Horner, G. (1999): Odour assessment with piezoelectric quartz crystal sensor arrays, a suitable tool for quality control in food technology? Adv. Food Sci. 21, 136-150.

Ohnishi, M., Ishibashi, T., Kijima, Y., Ishimoto, C., Seto, J. (1992): A molecular recognition system of odorants incorporating biomimetic gas-sensitive devices using Langmuir-Blodgett films. Sensors Mater. 4, 53-60.

O'Riordan, D.M.T., Wallace, G.G. (1986): Poly(Pyrrole-N-carbodithioate)electrode for electroanalysis. Anal. Chem. 58, 128-131.

Ozin, G.A. (1995): Microporous and mesoporous electronic materials: flexible open-framework nanomaterials for molecular recognition, towards the electronic nose. Supramol. Chem. 6, 125-134.

Pearce, T.C., Gardner, J.W., Friel, S., Bartlett, P.N., Blair, N. (1993) Electronic nose for monitoring the flavour of beers. Analyst, 118, 371-377.

Persaud, K., Dodd, G. (1982): Analysis of discrimination mechanisms in the mammalian olfactory system using a model nose. Nature 299, 352-355.

Schiffman, S.S., Leffingwell, J.C. (1981): Perception of odors of simple pyrazines by young and elderly subjects – a multidimensional-analysis. Pharmacol. Biochem. BE 14, 787-798.

Shurmer, H.V., Gardner, J.W., Chan, H.T. (1989): The application of discrimination techniques to alcohols and tobaccos using tin-oxide sensors. Sens. Actuators 18, 361-371.

Staples, E.J. (1998): Dioxin/Furane detection and analysis using a SAW based electronic nose. Presented at the 1998 IEEE Ultrasonics Symposium, Sendai,

78

Japan, Oct. 1998.

Staples, E.J. (1999): Electronic nose simulation of olfactory response containing 500 Orthogonal sensors in 10 seconds. Presented at the IEEE Ultrasonics Symposium, Lake Tahoe, CA, Oct. 18-21.

Vögtle, F., Hoss, R. (1992): Synthetic molecular receptor for piperazine and related diamines. J. Chem. Soc. Chem. Commun. 21, 1584-1585.

Von Loesecke, H.W. (1950): Bananas: Chemistry, Physiology, Technology. New York, Interscience.

Waltman, R.J., Bargon, J., Diaz, A.F. (1983): Electrochemical studies of some conducting polythiophene films. J. Phys. Chem. 87, 1459-1463.

Weimar, U., Schierbaum, K.D., Göpel, W. (1990): Pattern recognition methods for gas mixture analysis: Application to sensor arrays based upon SnO_2. Sens. Actuators B1, 93-96.

Winquist, F., Hornsten, E.G., Sundgren, H., Lündström, I. (1993): Performance of an electronic nose for quality estimation of ground meat. Meas. Sci. Technol. 4, 1493-1500.

Zhang, T., Petelenz, D., Janata, J. (1993): Temperature-controlled Kelvin Microprobe. Sens. Act. B12, 175-180.

4 RECEPTORS AS ANALYTICAL TOOLS FOR BIORESPONSE-LINKED INSTRUMENTAL ANALYSIS

U. Obst[1], G. Brenner-Weiß[1], M. Seifert[2], H. Sauerwein, E. Bauer, H.H.D. Meyer, B. Hock[2]

[1]Stadtwerke Mainz AG, Abtlg. Wasserversorgung, Rheinallee 41, 55118 Mainz, Germany
[2]Department of Botany, Technical University of Munich at Weihenstephan, Alte Akademie 12, D-85350 Freising, Germany
[3]University of Bonn, Institute of Anatomy, Physiology and Hygienics of Domestic Animals, Katzenburgweg 7-9, D-53115 Bonn, Germany
[4]Technical University of Muenchen at Weihenstephan, Department of Physiology, D-85350 Freising, Germany

Abstract. Receptor assays for estrogenic and androgenic compounds are presented as examples for bioresponse-linked analyses as receptor affinities are generally related to the strength of pharmacological effects. Since mixtures of different ligands yield additive signals, receptor assays can be conveniently used for effect monitoring, especially for screening programs. A detection limit with an enzyme-linked receptor assay (ELRA), based on the human estrogen receptor α, was achieved for 17-β-estradiol below 0.1 µg/l. A radioreceptor assay with the human androgen receptor reached a detection limit below 1 µg/l for dihydrotestosterone. The hyphenation of receptor affinity extraction with LC/MS/MS techniques was demonstrated for estrogens. It is considered a first step in bioresponse-linked instrumental analysis.

4.1   Introduction

Endocrine disruptors (ED) interact with the endocrine system and may therefore cause adverse effects on wildlife and human health, not only on an individual basis, but also on the population and community (Colborn 1995). ED encompass a variety of chemicals, including natural and synthetic hormones, natural plant constituents, pesticides, monomers and additives used in the plastics industry, detergent components and breakdown products, as well as persistent environmental pollutants (Soto et al. 1995). There is solid evidence for adverse effects of these chemicals in wildlife populations which have been exposed to high concentrations (Sumpter 1998). Exposure to a range of phytoestrogens, plant derived chemicals, is also of concern in some exposure scenarios (Hock and Seifert 1999).

The wide variety of known and unknown ED requires highly efficient methods for detecting relevant pollutants. The most effective approach is the exploitation of the respective receptors which are not only the recognition sites for hormones but also the binding sites of ED. Therefore it is possible to obtain effect-related signals with receptor assays, as documented for estrogenic and androgenic ingredients in illicit anabolic preparations (Rapp and Meyer 1987). However, they do not provide information on the chemical structure of the bound ED. Bioeffect-linked analyses are expected to solve this problem by combining biomolecular recognition processes with chemical analysis. It reduces costs of instrumental chemical analysis to relevant samples. The hyphenation of both steps provides' information on potential bioeffects, structure and concentration of the substances of interest. Presently it is assumed that a great number of environmental pollutants, which interfere with estrogenic and androgenic signal transduction chains, exert their effects by binding to estrogen and androgen receptors (Sohoni and Sumpter 1998) and can be used as biomolecular recognition elements.

4.2   Estrogen receptor

Several compounds which were originally considered harmless have been suggested to exert estrogenic effects, among them chemicals which are estrogen mimics (xenoestrogens), e.g. plastic softeners (bisphenol A) and detergents (4-nonylphenol). The wide variety of xenoestrogenic compounds makes it difficult to estimate the toxic potential from their structures. Thus, environmental monitoring cannot be performed by chemical analysis of water ingredients for the presence of known or suspected structures. Test systems should also be able to detect compounds which were previously unknown as xenoestrogens. As the majority of estrogenic effects are mediated by estrogen receptors (Clark 1998) analytical tools based on these receptors can be established. Up to now two estrogen receptors ER) have been identified: ERα (Meyer and Jungblut 1983, Green et al. 1986) and ERβ (Mosselmann et al. 1996). Both receptors share the same basic structure, but have different affinities for several estrogens and xenoestrogens. Especially phytoestrogens show higher affinities to the ERβ (Kuiper et al. 1997). However, most test systems presently use the ERα.

Different approaches for detecting estrogenic effects in environmental samples have been applied. They can be classified into three different groups: cell culture-based assays, reporter gene assays and subcellular receptor binding assays.

Cell culture-based assays determine the estrogen-dependent induction of cell proliferation in estrogen sensitive tumor cells like MCF-7. The E-SCREEN developed by Soto et al. (1995) is based on the cell yield achieved after a 4-6 day incubation in the presence or absence of estrogens, respectively.

There are two types of reporter gene assay test sytems with (1) cells, with genuine expression of ER (e.g. MCF-7 cells), and (2) cells genetically modified for expression of ER. In the first case cells are used, which have been transfected with a reporter gene (Pons et al. 1990, Gagne et al. 1994), which is measurably induced after estrogen exposure. In the second case, double transfection is needed. The cells (mainly yeasts) have to be equipped with a gene for ER expression and the reporter gene construct (Arnold et al. 1996). With yeast cells, problems of membrane permeability and transport have been identified that may affect the measurement of the relative estrogenic potential (Zysk et al. 1995).

Less complex approaches to detect estrogenic substances are the subcellular receptor binding assays. Receptors can be applied as binding proteins for pharmacologically and toxicologically relevant substances. Cell extracts containing natural ER derived from uteri of several animal species as well as genetically engineered ER can be used. Several recombinant expression systems based on yeast cells (McDonnell et al. 1991) or baculovirus transfected insect cells (Brown and Sharp 1990) are known. Up to now mostly radioreceptor assays have been used. They are based on radioactively labelled 17ß-estradiol (Korenman 1968). The parameter measured is the relative binding affinity compared to 17ß-estradiol. Results are expressed as IC_{50} values or as relative binding affinities (RBA) compared to 17ß-estradiol.

Although ER-based radioreceptor assays have been available for many years, they were not suited for routine applications in the environmental field because of their dependence on radioactive tracers. Consequently the development of non-radioactive assays for estrogens and xenoestrogens is of major interest. Available assays are based on different detection principles. Oosterkamp et al. (1996) used the on-line coupling of reversed-phase liquid chromatography and biochemical detection based on receptor-ligand interactions for discovering estrogenic compounds in urine. This receptor affinity detection (RAD) system allowed the simultanous measurement of different estrogenic substances within a few minutes.

Another non-radioactive receptor assay applied fluorescence polarisation, which is detected with a polarimeter (Bolger et al. 1998).

For the detection of estrogens and xenoestrogens in environmental samples an Enzyme-Linked Receptor Assay (ELRA) in microwell format was recently developed by Seifert et al. (1999). This receptor assay uses the same principles as an indirect competitive enzyme-immunoassay based on ligand-protein interactions, but it employs the full length recombinant human ERα instead of antibodies as specific binding proteins. The ELRA measures the competition of sample estrogen and 17ß-estradiol supplied as a BSA-coating conjugate for the binding site of the dissolved ER. The individual steps are shown in Fig. 4.1. After removing the non-bound receptor by a washing step, the bound receptors form an immunocomplex with a biotinylated anti-ER antibody directed against the DNA-binding site of the receptor. Followed by a further washing step, a streptavidin-peroxidase-biotin complex is added. After washing the peroxidase substrates tetramethylbenzidine and H_2O_2 are incubated. Substrate turnover is stopped, and the absorption is measured using a microtiter plate reader. A detection limit of 0.1 µg/l for 17ß-estradiol can be achieved with the chromogene substrate tetramethylbenzidine. A detection limit of 0.02 µg/l could be achieved using the luminescent substrate luminol (Seifert, unpublished). Similar to the ELISA, the ELRA shows an inverse relation between the free estrogen concentration and the color intensity.

Relative binding affinities of estrogenic and non-estrogenic substances were determined with the ELRA by comparing the middle of the test curve (IC_{50}; µg/l) of the cross-reacting substances with 17ß-estradiol. Fig. 4.2 shows an ELRA calibration curve using the luminescent substrate luminol. Substances with known estrogenic potential were assayed and include 17ß-estradiol metabolites, synthetic chemicals used as contraceptives (e.g. ethinylestradiol) and for medical purposes (e.g. tamoxifen) as well as industrial chemicals (e.g. 4-nonylphenol, bisphenol-A).

84

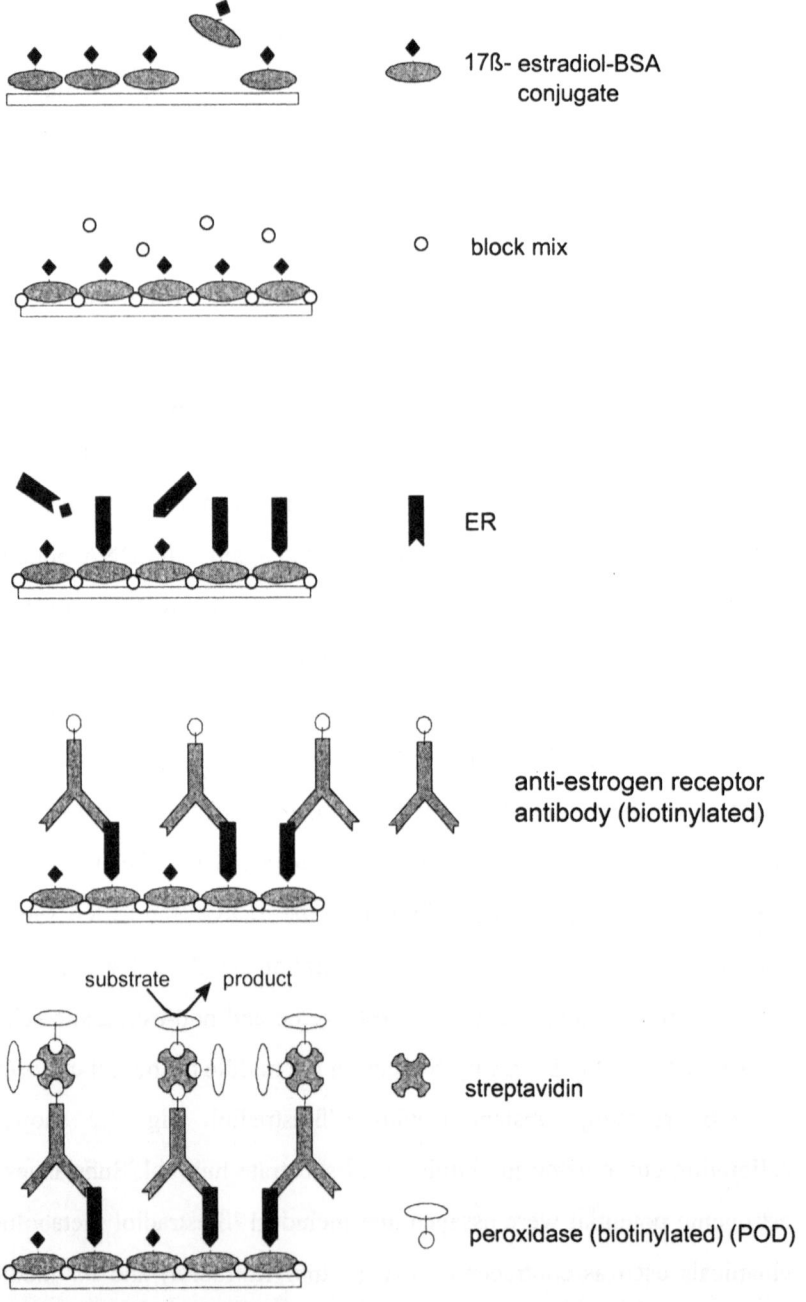

Fig. 4.1: Test principle of ELRA

Substances with chemical structures comparable to 17ß-estradiol but lacking estrogenic activity such as steroids like testosterone or cortisone were also included as negative control. The ELRA results obtained for known estrogenic and xenoestrogenic substances closely matched radioreceptor assay data (Kuiper et al. 1997). In addition, the ELRA proved suitable for measuring real water samples. Hence the ELRA is a sensitive non-radioactive receptor assay for cost effective, high through-put screening of environmental samples. Mixtures of different estrogens and xenoestrogens always revealed additivity.

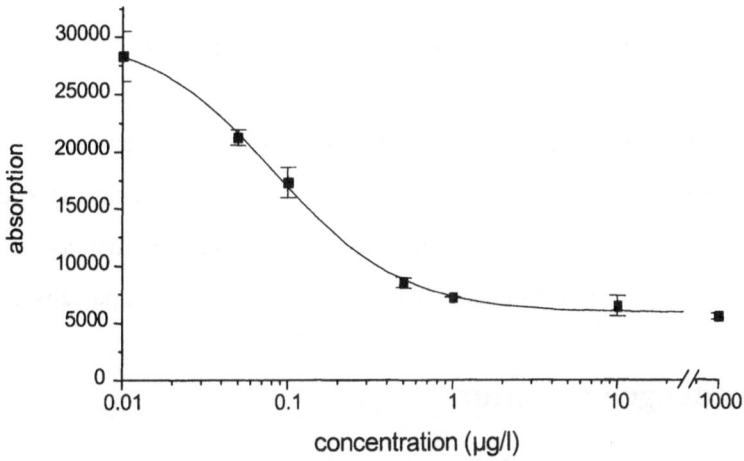

Fig. 4.2: ELRA calibration curves using the luminescent substrate luminol

Since the ELRA is able to detect all substances which bind to the receptor, it is an bioeffect-linked test system. However, it is not possible to distinguish agonists from antagonists, as these receptor binding tests do not provide information whether a ligand activates or blocks the receptor.

Garrett at al. (1999) extended non–radioactive receptor assay technology in microwell format to food analysis. In this test system the natural estrogen 17ß-estradiol is used as a tracer. Ligands in the sample compete with 17ß-estradiol for binding to the ER. Unbound 17ß-estradiol is subsequently measured by an immunoassay. As the core element of this assay is the recombinant hormone binding domain of the human ERα, the transcriptional activity of a ligand, similar to the systems mentioned above, can not be detected.

Cheskis et al. (1997) measured the kinetics of human ER binding to the estrogen response element (ERE) using a BIAcore biosensor system. DNA-binding was analysed on-line in the absence and presence of several estrogenic and anti-estrogenic compounds. They also showed that there is good correlation between the kinetics of ER-ERE interaction induced by a biological ligand and its biological effect.

Since the strength of the receptor binding assay is seen in its application as a screening test, positive samples can be re-analysed with cellular assays to specify agonistic or antagonistic effects. In addition, analytical techniques such as LC-MS can be applied for subsequent chemical identification of receptor-bound substances.

4.3   Androgen receptor

In addition to those environmental contaminants that interfere with the ER, there is a number of compounds, which are suspected for androgenic or anti-androgenic activities. Some of the environmental chemicals exert androgenic and anti-estrogenic actions not via direct interaction with the androgen receptor (AR), but by inhibiting one of the key enzymes of androgen synthesis, i.e. by suppressing aromatase cytochrome P450 which converts C-19 androgens to aromatic C-18 estrogenic steroids. This type of interference has been suggested for tributyltin (Bettin et al. 1996) and two flavones (Mak et al. 1999).

Those substances which are known to interfere with androgen effects at the receptor level, i.e. via binding to the AR, mainly include various natural and synthetic pesticides, e.g. pyrethroids, vinclozolin, linuron and diuron (Eil and Nisula 1990, Gray et al. 1994, Cook et al. 1993, Bauer et al. 1998). Anti-androgenic activity has also been reported for a derivative of the insecticide DDT, the persistent metabolite p,p'-DDE (Kelce et al. 1994).

For the identification of substances which bind to the AR, similar analytical approaches are available as described for estrogenic activities. Reporter gene-based approaches with transgenic cell lines have been reported for AR activation or blockade (Vinsgaard et al. 1999, Mak et al. 1999, Fuhrmann et al. 1992). Recently, a yeast assay system has been published in which both targets of androgen interference are combined: aromatase and AR are simultaneously used for screening environmental chemicals for their anti-aromatase activity as well as for their interaction with the AR (Mak et al. 1999). With respect to bioeffect-linked analyses, analytical approaches are preferred which use cell-free systems, where the target molecules enable recognition. AR, like ER, is available from animal tissues, and for decades radioreceptor assays using tissue preparations are well established techniques to characterize therapeutics, anabolics or environmental pollutants in terms of their mode of action (e.g. Danhaive and Rousseau 1986, Kelce et al. 1994, Bauer et al. 1998). Preparations from typical androgen target tissues, such as accessory glands of the male reproductive tract derived from individuals with low endogenous sex hormone secretion (juvenile or castrated), have sufficiently high concentrations of free AR, i.e. AR which is not withheld in the nucleus due to prior ligand activation.

To evaluate the relative binding affinities of different ligands to the AR, competitive radioreceptor assays are usually performed. The natural ligands of the AR, dihydrotestosterone (DHT) and testosterone, are used in the tritiated form as

AR labels. To prevent errors arising from label degradation during incubation, synthetic ligands may be used alternatively. After incubating the cytosol with a constant concentration of the radiolabeled ligand (tracer) and the unknown concentration of the analytes in the sample, bound and free ligands are separated. Separation techniques include treatment with dextrane-coated charcoal or hydroxylapatite. Bound ^{3}H-labelled steroids are subsequently quantified. The concentrations of the substances, which inhibit 50% binding of the radiolabeled ligand are then derived from the displacement curves. An example for a radioreceptor assay is given in Fig. 4.3.

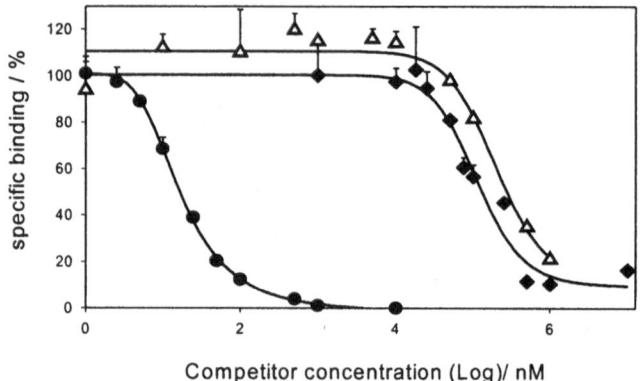

Fig. 4.3: Displacement curves showing the relative abilities of (●) DHT(υ) 3,4-DCAc and (Δ) 3,4-DCA to compete with binding in calf uterus cytosol. Data points are means ± s from 4-8 replicates assayed in 2-4 tests. The curves depicted were calculated according to the four parameter logistic function

It shows the potency of different substances to displace [^{3}H]-DHT from the bovine AR (Bauer et al. 1998). A comparison of the displacement curves obtained with 3,4-dichloroaniline (3,4-DCA) and its metabolite 3,4-dichloroacetanilide (3,4-DCAc) showed the higher affinity of 3,4-DCAc to the AR. 3,4-DCA is a phenylurea herbicide which also occurs in waste products of dye production. It is rapidly taken up by fish and metabolized to 3,4-DCAc. These data indicate that the endocrine disrupting activity of individual pollutants can be altered by metabolization of the respective substance in the exposed species.

Receptor preparations from tissues have several disadvantages. Receptor contents vary, therefore standard preparations with reproducible receptor concentrations can not be provided for routine analyses. In addition, tissues usually not only contain one receptor type, but several other types. For example, in uterine tissue several different steroid hormone receptors are present, e.g. ER, AR, gestagen plus glucocorticoid receptor. When synthetic derivatives of the natural ligands are used, special care has to be taken to determine first the receptor binding specificities of these ligands, since several ligands are known to bind also to other steroid receptors. For instance methyltrienolone and mibolerone which are frequently used to label AR, also bind to the gestagen receptor (Murthy et al. 1986). To circumvent this problem, the cross-reacting receptors might be saturated with a special unlabeled ligand, as described earlier (Sauerwein and Meyer 1989). Taking these limitations into account, recombinant AR protein is preferable. Until now, however, no cell-free assay systems have been available, in which recombinant AR is used to screen environmental samples. Independent from the source of AR, species-specificity of AR recognition has to be considered since differences in ligand affinities are known even within mammalia at least for gestagen receptors (Jewgenow and Meyer 1998). There are several reports indicating that sex steroid hormone receptors from species living in aquatic environments differ from those of man in terms of affinity and specificity (Fitzpatrick et al. 1994, Le Drean et al. 1995, Sperry and Thomas 1999). This is of general relevance in all bioeffects-linked analytical systems; therefore the target species have to be clearly defined.

The successful production of human androgen receptor (hAR) was first reported by Chang et al. (1992) for baculovirus infected insect cells. Using a similar system, Bauer et al. (2000) generated recombinant hAR and developed a receptor assay in which the AR is immobilised via specific antibodies in microtiter plates. The hAR binding characteristics established in this assay format are in agreement with those reported in the literature. The assay system has been tested for its applicability in testing environmental samples. Fig. 4.4 demonstrates that the standard curves are neither shifted nor distorted in presence of water samples from various sources.

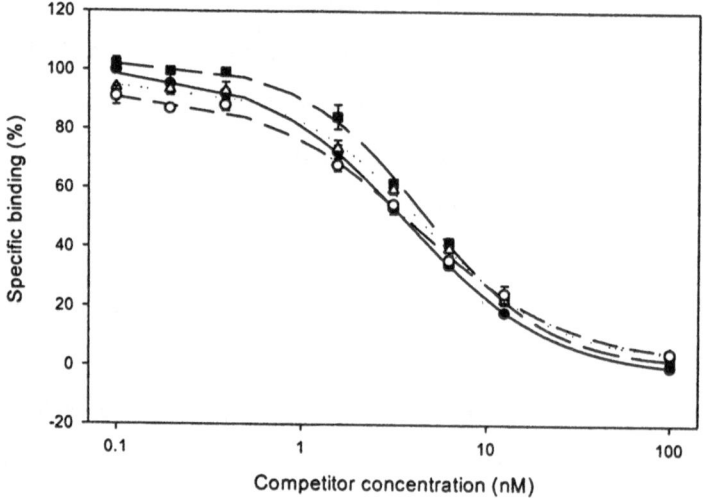

Fig. 4.4: Displacement curves showing the effect of dihydrotestosterone spikes (0.2 nM -100 nM) in ultrapure water (Δ), tap water ($\bullet$), river water ($\circ$) and sewage effluent ($\blacksquare$) in displacement of [^{3}H]-dihydrotestosterone bound to the androgen receptor. Data points are means ± s from 3 replicates

In conclusion, the assay is sufficiently robust to screen environmental samples for the presence of substances with potential AR binding activity. The limit of detection is 1 ng/mL DHT. Extracting water samples prior to the receptor assay, e.g. by hydrophobic solid phase columns, is expected to improve the sensitivity but may also imply a certain structural pre-selectivity leading to a potential exclusion of substances, which are not retained on the column. In view of the analytical potential for integrating recombinant AR as recognition molecule in chemical sample treatment, fractionation and identification, the aspect of bioeffect-linked approaches can further be extended to those substances which might interfere in androgen action. Moreover, different receptors, e.g. AR and ER, might be applied simultaneously in such analytical systems and thus increase the information available from individual samples without increasing the number of tests.

4.4 Chemical analytics of estrogens and androgens

Because of the structural diversity of estrogenic or androgenic active substances a variety of analytical methods has to be used for the determination of estrogens and androgens. Usually, immunochemical procedures like RIA or ELISA are sensitive but fail to reveal the structural confirmation of these substances. UV and fluorescence detection are not specific enough for the identification of substances of similar structure or substances in complex matrices. Therefore the importance of mass spectrometry (MS) grew due to its high specifity, sensitivity and ability to provide structural information. Particularly in combination with powerful separation techniques like gas chromatography (GC) or high performance liquid chromatography (HPLC), MS has proven to be a useful tool in chemical analysis of complex mixtures. GC/MS for example is a well-established technique in environmental analytics, but has the disadvantage that only volatiles can effectively be separated. This means that GC techniques fail to separate thermolabile and high molecular mass compounds and that substances with polar functional groups (e.g. amino- or hydroxy-groups etc.) have to be derivatized (Knapp 1979). Figure 4.5 shows an exemplary chromatogram of an estrogenic standard mixture in water determined by GC/MS (Stumpf et al. 1996). After solid phase extraction (SPE) and silica gel clean-up the steroids have to be silylated before GC/MS detection.

92

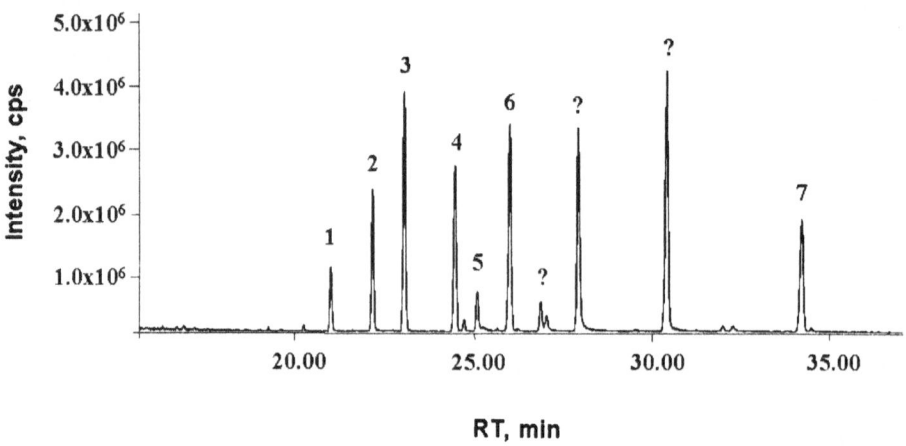

Fig. 4.5: GC/MS-Chromatogramm of selected steroid hormones (original data from Stumpf et al. 1996). Substances were silylated before GC/MS detection. 1 = Mirex (internal standard, 2 = estrone, 3 = estradiol, 4 = mestranol, 5 = estradiol acetate, 6 = ethinylestradiol, 7 = estradiol valerate, ? = unknown

For the separation of polar or thermolabile compounds LC techniques should be preferred because derivatization steps can usually be avoided and the sample pretreatment is simplified. Since several modes of separation are available in modern HPLC techniques (e.g. reverse phase chromatography (RPC), normal polar chromatography (NPC), size exclusion chromatography (SEC) etc.), analytes of different chemical properties as well as high molecular compounds can be separated. Currently, the most popular approaches to LC/MS involve electrospray ionization (ESI) and atmospheric pressure chemical ionization (APCI) (Fenn 1993, Jahn and Jahn 1997). Both APCI and ESI allow the determination of substances in very low concentrations.

In the following, an example for environmental analysis of steroid hormones at a trace level is given using LC/MS/MS without derivatization of the substances to be assayed. For the analysis of estrogens and androgens in surface water and drinking water Brenner-Weiß et al. (1999) developed a method using solid phase extraction with RP-C$_{18}$ material followed by LC/MS/MS detection. Nine steroid hormones could simultaneously be analyzed in the range of a few ng/L. Recovery rates were about 80 to 100% in most cases within detection limits of 20 ng/L for each compound. Figure 4.6 shows the analysis of a steroid standard mixture using positive APCI conditions. The steroids were detected in the multiple reaction monitoring mode (MRM), by which single mass transitions characteristic for each substance were acquired.

4.5 Receptor affinity extraction-liquid chromatography tandem mass spectrometry (RAE-LC/MS/MS) as an example for bioeffects-related analytics of estrogens

Receptor assays such as ELRA indicate the biological activity of unknown substances, whereas LC/MS/MS techniques must be used for structural identification. Therefore the hyphenation of both methods is an excellent example for bioeffects-linked analytics. In the following a combination of receptor affinity extraction (RAE) with liquid chromatography tandem mass spectrometry is briefly described for the detection of estrogens in water. Receptor binding is used for separation and specific enrichment of estrogenic active substances as part of the receptor affinity extraction (RAE) (Brenner-Weiß et al. 1999). Because covalent immobilization of the ER causes a decrease in receptor activity and affinity(Ikeda et al. 1988), the ER is directly incubated with water samples containing estrogens. Then the receptor-ligand complex is trapped by a protamine-agarose conjugate. Protamine belongs to a group of basic binding proteins associated with DNA in sperm particulary of fish, birds etc. and is able to bind different steroid receptors in the uncharged and charged form (Chamness et al. 1975). Subsequently, the incubation solution is filtered through a conditioned cartridge filled with additional

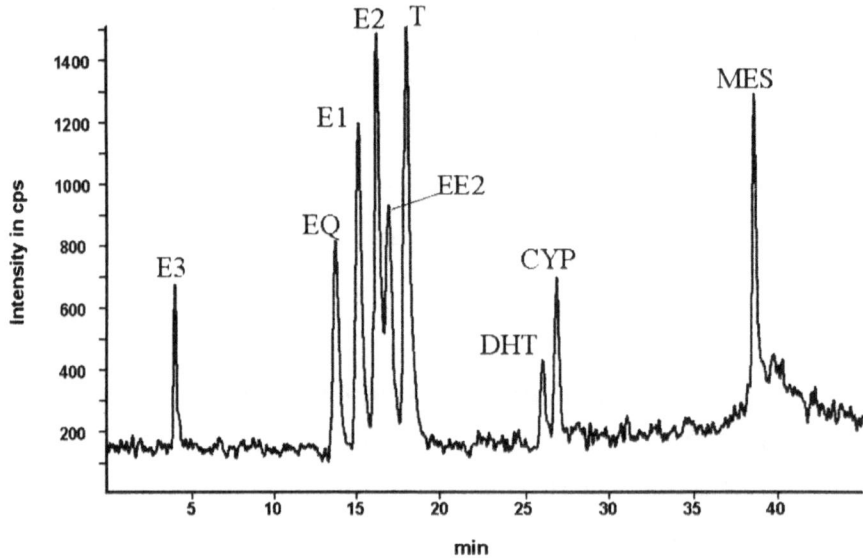

Fig. 4.6: LC-MS/MS chromatogram of a standard mixture of nine steroid hormones. Data are acquired in the MRM mode. E3 = estriol, EQ = equilin, E1 = estrone, E2 = estradiol, T = testosterone, EE2 = ethinylestradiol, DHT = dihydrotestosterone, CYP = cyproterone acetate, MES = mestranol

protamine-agarose gel. The procedure is similar to the assays described by van der Vlis et al. (1994) where protamine-coated glass fibre filters or silica gel were used. After release from the receptor by elution buffer, the estrogen was determined by LC/MSMS. A scheme of this procedure is shown in Figure 4.7.

The results shown in Figure 4.8 prove that 17β-estradiol can only be extracted and subsequently determined in the eluate when ER is added to the solution.

Figure 4.9 illustrates the analysis of river water spiked with 17β-estradiol as an example for bioeffects-linked analytics of estrogens in surface water. Estrogen concentrations in the lower ng/L range need a preconcentration of the sample by lyophilization. After lyophilization of the water sample 17β-estradiol can be determined in the range of 10ng/L.

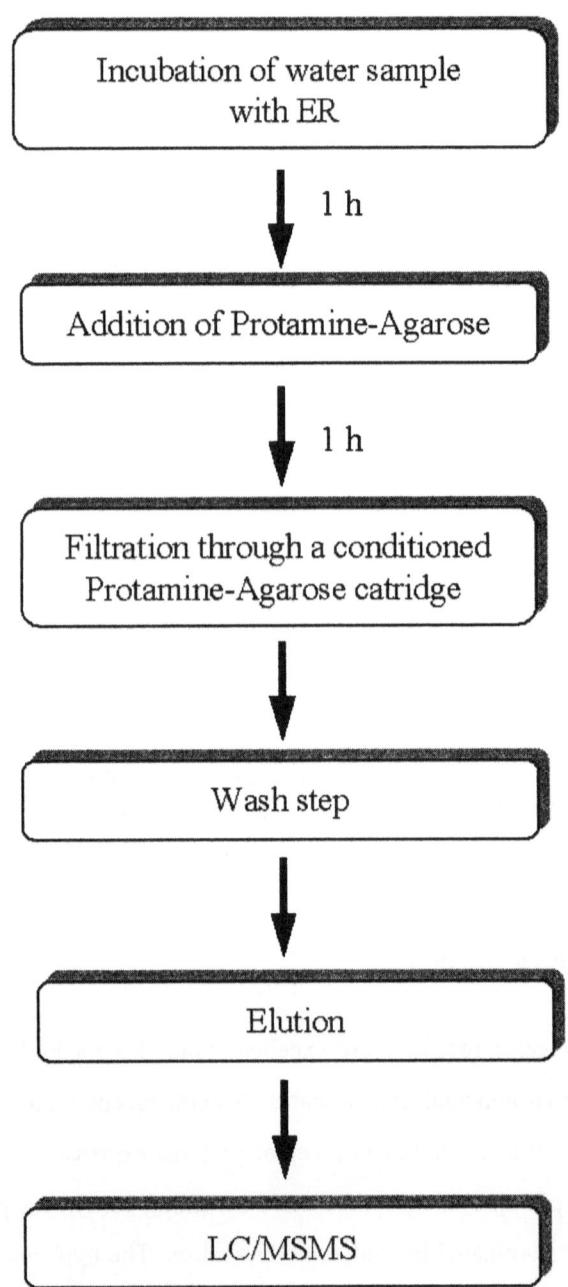

Fig. 4.7: Scheme of RAE-LC/MSMS

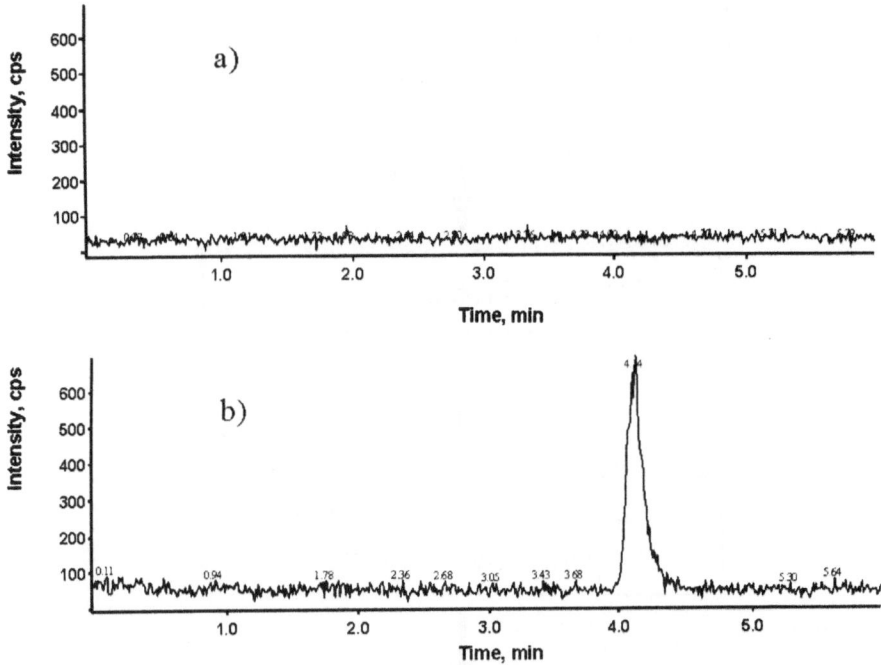

Fig. 4.8: RAE-LC/MSMS of standard solution with PA (original data from G. Brenner -Weiß); a) 1 ng 17β-estradiol in 1 mL PBS buffer without ER b) 1 ng 17β-estradiol in 1 mL PBS buffer with ER

4.6 Conclusions

Estrogen and androgen receptors are excellent examples for biological molecular recognition of environmental contaminants. Natural receptor molecules extracted from tissues as well as recombinant receptor proteins expressed in host cells were used as binding proteins in non-radioactive screening assays for pre-selection of water samples contaminated by endocrine disruptors. The hyphenation of receptor assays with mass spectrometry provides the first practical and successful examples for bioeffect-linked instrumental analysis. For environmental monitoring enzyme-

linked receptor assays (ELRA) were adapted to receptor affinity extraction (RAC) to get a bioeffect-linked selection and enrichment of potential endocrine disruptors.

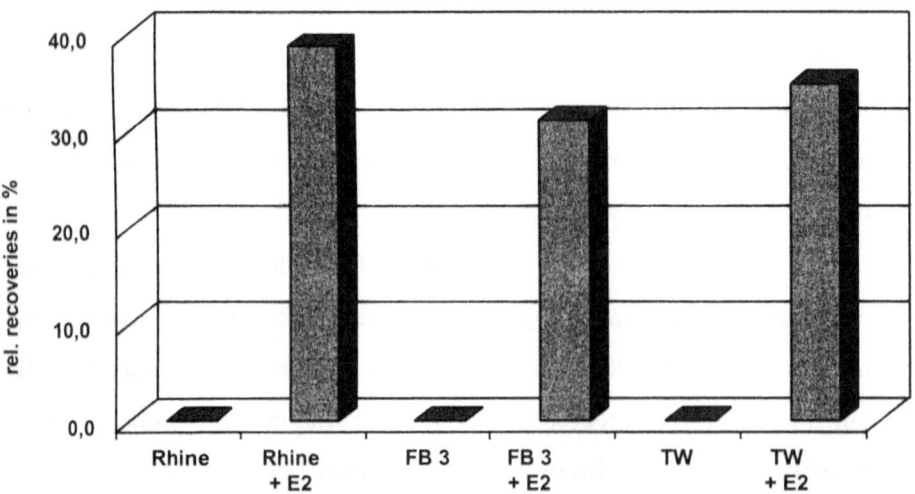

Fig. 4.9: RAE-LC/MSMS of real water samples (original data from G. Brenner-Weiß) with and without spiking (10 ng/L 17β-estradiol); river Rhine at Mainz, FB = embankment filtrate, TW = drinking water

After elution structure and concentration of the receptor-bound ligands were determined by LC/MS/MS. Receptor affinity extraction-liquid chromatography tandem mass spectrometry succeded in detecting estrogenic substances in river water and drinking water and is therefore the first step for the implementation of bioeffect-linked instrumental analysis in environmental monitoring.

Acknowledgements. We would like to thank the BMBF for the financial support (project 02 WV 994/0).

98

4.7 References

Arnold, S.F., Robinson, M.K., Notides, A.C., Guillette, L.J., McLachlan, J.A. (1996): A yeast estrogen screen for examining the relative exposure of cells to natural and xenoestrogens. Environ. Health Perspect. 104, 544-548.

Bauer, E.R.S., Meyer, H.H.D., Stahlschmidt-Allner, P., Sauerwein, H. (1998): Application of an androgen receptor assay for the characterisation of the androgenic or antiandrogenic activity of various phenylurea herbicides and their derivatives. Analyst 123, 2485-2487.

Bauer, E.R.S., Sauerwein, H., Petri, T., Meyer, H.H.D. (2000): Efficient expression of recombinant human androgen (hAR) receptor form baculovirus infected insect cells and development of a microtiter plate receptor assay for AR-binding chemicals. Proceedings of the EuroResidue IV, Conference on Residues of Veterinary Drugs in Food, Veldhoven The Netherlands.

Bettin, C., Oehlmann, J., Stroben, E. (1996): TBT-induced imposex in marine neogastropods is mediated by an increasing androgen level. Helgol. Meeresunters. 50, 299-317.

Bolger, R., Wiese, T.E., Ervin, K., Nestich, S., Checovich, W. (1998): Rapid screening of environmental chemicals for estrogen receptor binding capacity. Environ. Health Perspect. 106, 551-557.

Brenner-Weiß, G., Nusser, M., Kirschhöfer, F., Obst, U. (1999): Wirkungsbezogene Analytik von Umweltschadstoffen mit Rezeptortests; Teilprojekt 4: Chemische Validierung des Tests und Identifizierung der Rezeptorwirksamen Substanzen. BMBF-Forschungsvorhaben 02 WU 9650/1.

Brown, M., Sharp, P.A. (1990): Human estrogen receptor forms multiple protein-DNA complexes. J. Biol. Chem. 265, 11238-11243.

Chamness, G.C., Huff, K., McGuire, W.L. (1975): Protamine-precipitated estrogen receptor: A solid phase ligand exchange assay. Steroids 25, 627–635.

Chang, C., Wang, C., DeLuca, H.F., Ross, T.K., Shih, C.C.Y. (1992): Characterisation of human androgen receptor overexpressed in the baculovirus system. Proc. Natl. Acad. Sci. USA 89, 5946-5950.

Cheskis, B.J., Karathanasis, S., Lyttle, C.R. (1997): Estrogen receptor ligands modulate its interaction with DNA. J. Biol. Chem. 272, 11384-11391.

Clark, J. (1998): Female reproduction and toxicology of estrogen, pp. 259-275. In: Reproductive and developmental toxicology (Korach, K., ed.). Marcel Dekker, New York.

Colborn, T. (1995): Environmental estrogens: Health implications for humans and wildlife. Environ. Health Perspect. 103 (Suppl 7), 135-136.

Cook, J.C., Mullin, L.S., Frame, S.R., Biegel, L.B. (1993): Investigation of a mechanism for ledying cell tumorigenesis by linuron in rats. Toxicol. Appl. Pharmacol. 119, 195-204.

Danhaive, P.A., Rousseau, G.G. (1986): Binding of glucocorticoid antagonists to androgen and glucocorticoid hormone receptors in rat skeletal muscle. J. Steroid Biochem. 24, 481-487.

Eil, C., Nisula, B.C. (1990): The binding properties of Pyrethroids to human skin fibroblast androgen receptors and to sex hormone binding globulin. J. Steroid Biochem. 35, 409-414.

Fenn, J.B. (1993): Ion formation from charged droplets: Roles of geometry, energy, and time. J. Am. Soc. Mass Spectrom. 4, 524.

Fitzpatrick, M.S., Gale, W.L., Schreck, C.B. (1994): Binding characteristics of an androgen receptor in the ovaries of coho salmon, *Oncorhynchus kisutch*. Gen. Comp. Endocrinol. 95, 399-408.

Fuhrmann, U., Bengtson, C., Repenthin, G., Schillinger, E. (1992): Stable transfection of androgen receptor and MMTV-CAT into mammalian cells: inhibition of cat expression by anti-androgens. J. Steroid Biochem. Mol. Biol. 42(8),787-793.

Gagne, D., Balaguer, P., Demirpence, E., Chabret, C., Trousse, F., Nicolas, J. (1994): Stable luciferase transfected cells for studying steroid receptor biological activity. J. Biolumin. Chemilumin. 9, 201-207.

Garrett, S.D., Lee, H.A., Morgan, M.R. (1999): A nonisotopic estrogen receptor-based assay to detect estrogenic compounds. Nat. Biotechnol. 17, 1219-22.

Gray, L.E., Jr., Ostby, J.S., Kelce, W.R. (1994): Developmental effects of an environmental antiandrogen: The fungicide vinclozolin alters sex differentiation of the male rat. Toxicol. Appl. Pharmacol 129, 46-52.

Green, S., Walter, P., Kumar, V., Krust, A., Bornert, J.-M., Argos, P., Chambon, P. (1986): Human estrogen receptor cDNA: Sequence, expression and homology to v-erb-A. Nature 320, 134-139.

Hock, B., Seifert, M. (1999): Detecting estrogens in food and the environment. Nat. Biotechnol. 17, 1162-1163.

Ikeda, M., Watanabe, S., Kusaka, T. (1988): Covalent immobilization of the estrogen receptor to a cationic N-hydroxysuccinimide ester derivative of agarose. Steroids 52(3), 187-203.

Jahn, K., Jahn, K. (1997): Massenspektrometrie – Kopplungstechniken mit der chromatographie. Nachr. Chem. Tech. Lab. 45, M1–M9.

Jewgenow, K., Meyer, H.H.D. (1998): Comparative Binding Affinity Study of Progestins to the Cytosol Progestin Receptor of Endometrium in Different Mammals. Gen. Comp. Endocrinol. 110, 118-124.

Kelce, W.R., Monosson, E., Gamcsik, M.P., Laws, S.C., Gray, L.E. (1994): Environmental hormone disruptors: Evidence that vinclozolin developmental toxicity is mediated by antiandrogenic metabolites. Toxicol. Appl.

Pharmacol. 126, 276-285.

Knapp, D.R. (1979): Handbook of analytical derivatization reactions. Wiley & Sons, London.

Korenman, S.G. (1968): Radio-ligand binding assay of specific estrogens using a soluble uterine macromolecule. J. Clin. Endo. Metab. 28, 127-130.

Kuiper, G.G.J.M., Carlsson, B., Grandien, K., Enmark, E., Häggblad, J., Nilsson, S., Gustafsson, J.A. (1997): Comparison of the ligand binding in specifity and transcript tissue distribution of estrogen receptors alpha and beta. Endocrinology 138, 863-870.

Le Drean, Y., Kern, L., Pakdel, F., Valotaire, Y. (1995): Rainbow trout estrogen receptor presents an equal specificity but a differential sensitivity for estrogens than human estrogen receptor. Mol. Cell. Endocrinol. 109, 27-35.

Mak, P., Cruz, F.D., Chen, S. (1999): A yeast screen system for aromatase inhibitors and ligands for androgen receptor: Yeast cells transformed with aromatase and androgen receptor. Environ. Health Perspect. 107, 855-860.

McDonnell, D.P., Nawaz, Z., Densmore, C., Weigel, N.L., Pham, T.A., Clark, J.H., O'Malley, B.W. (1991): High level expression of biologically active estrogen receptor in *Saccharomyces cerevisiae*. J. Steroid Biochem. Molec. Biol. 39, 291-297.

Meyer, H.H.D., Jungblut, P.W. (1983): Purification of the steroid-binding core of porcine estrogen receptor. Hoppe Seyler's Z. Physiol. Chemie 364, 757-768.

Mosselman, S., Polman, J., Dijkema, R. (1996): ERbeta: Identification and characterization of a novel human estrogen receptor. Fed. Europ. Biochem. Soc. 392, 49-53.

Murthy, L.R., Johnson, M.P., Rowley, D.R., Young, C.F.Y., Scardino, P.T., Tindall, D.J. (1986): Characterisation of Steroid Receptors in Human Prostate Using Mibolerone. Prostate 8, 241-253.

Oosterkamp, A.J., Villaverde-Herraiz, M.T., Irth, H., Tjaden, U.R., van der Greef, J. (1996): Reversed-phase liquid chromatography coupled on-line to receptor affinity detection based on the human estrogen receptor. Anal.Chem. 68, 1201-1206.

Pons, M., Gagne, D., Nicolas, J., Mehtali, M. (1990): A new cellular model of response to estrogens: a bioluminescent test to characterize (anti) estrogen molecules. Biotechniques 9, 450-459.

Rapp, M., Meyer, H.H.D. (1987): Untersuchung von illegalen, nicht deklarierten Injektionspräparaten auf sexualhormonwirksame steroidale Anabolika und Entwicklung eines Radioimmuntests für 19-Nortestosteron. Arch. F. Lebensmittelhyg. 38, 35-41.

Sauerwein, H., Meyer, H.H.D. (1989): Androgen and estrogen receptors in bovine skeletal muscle: relation to steroid-induced allometric muscle growth. J. Anim. Sci. 67, 206-12.

Seifert, M., Haindl, S., Hock, B. (1999): Development of an enzyme linked receptor assay (ELRA) for estrogens and xenoestrogens. Anal. Chim. Acta 386, 191-199.

Sohoni, P., Sumpter, J.P. (1998): Several environmental oestrogens are also anti-androgens. J. Endocrinol. 58, 327-339.

Soto, A.M., Chung, C., Fernandez, M.F., Olea, N., Olea-Serrano, M.F. (1995): The E-Screen assay as a tool to identify estrogens: An update on estrogenic environmental pollutants. Environ. Health Perspect. 103, 113-119.

Sperry, T.S., Thomas, P. (1999): Identification of two nuclear androgen receptors in kelp bass (*Paralabrax clathratus*) and their binding affinities for xenobiotics: Comparison with Atlantic croaker (*Micropogonias undulatus*) androgen receptors. Biol. Reprod. 61, 1152-1161.

Stumpf, M., Ternes, T., Haberer, K., Baumann, W. (1996): Nachweis von natürlichen Östrogenen in Kläranlagen und Fließgewässern. Vom Wasser 87, 251–261.

Sumpter, J.P. (1998): Xenoendorine disrupters--environmental impacts. Toxicol. Lett. 102-103, 337-42.

Van der Vlis, E., Irth, H., Tjaden, U.R., van der Greef, J. (1994): Potential or receptor-ligand interactions for sample handling in liquid and gas chromatography. J. Chromatogr. A 665, 233-241.

Vinsgaard, A.M., Joergensen, E.C., Larsen, J.C. (1999): Rapid and sensitive reporter gene assays for detection of antiandrogenic and estrogenic effects of environmental chemicals. Toxicol. Appl. Pharmacol. 155, 150-60.

Zysk, J.R., Johnson, B., Ozenberger, B., Bingham, B., Gorsky, J. (1995): Selective uptake of estrogenic compounds by *Saccharomyces cerevisiae*: A mechanism for antiestrogen resistance in yeast expressing the mammalian estrogen receptor. Endocrinology 136, 1323-1331.

5 BIORESPONSE-LINKED ANALYSIS BASED ON ACETYLCHOLINESTERASE INHIBITION

Till Bachmann, Jürgen Pleiss, Francois Villatte, Rolf D. Schmid

Institute for Technical Biochemistry, University of Stuttgart, Allmandring 31, 70569 Stuttgart, Germany

Abstract. Acetylcholinesterase is a key enzyme in the nervous system. This protein is the biological target of the predominant insecticides used in agriculture and other neurotoxins such as algal toxins. Its high sensitivity, rapid turn-over, and stability promoted its use in analytical applications to detect pesticides at low concentrations in environmental samples. It is possible today to produce this enzyme recombinantly in different systems and to engineer mutated forms based on structural information and computer modeling techniques leading to a large variety of enzyme properties. In this chapter the main features of acetylcholinesterase with regard to its biotechnological use and the last developments in biosensors are presented.

5.1 Introduction

Acetylcholinesterases (AChE) are key enzymes involved in the transmission of neurosignals in higher animals. They function by removing acetylcholine from the synaptic space between neurons after the nervous impulse thus allowing the repolarization of the post-synaptic membrane. Inhibition of AChE interferes with this process and may lead to a severe impairment of nerve functions or even to death. Organophosphates such as „nerve gases" and some algal toxins are strong inhibitors of AChE. As organophosphates and carbamates may also act as inhibitors of insect AChE, they are widely used as contact insecticides. AChE is among the most active enzymes presently known. The turnover number of human AChE is very high, in the order of 25,000 s^{-1}, indicating diffusion limitation.

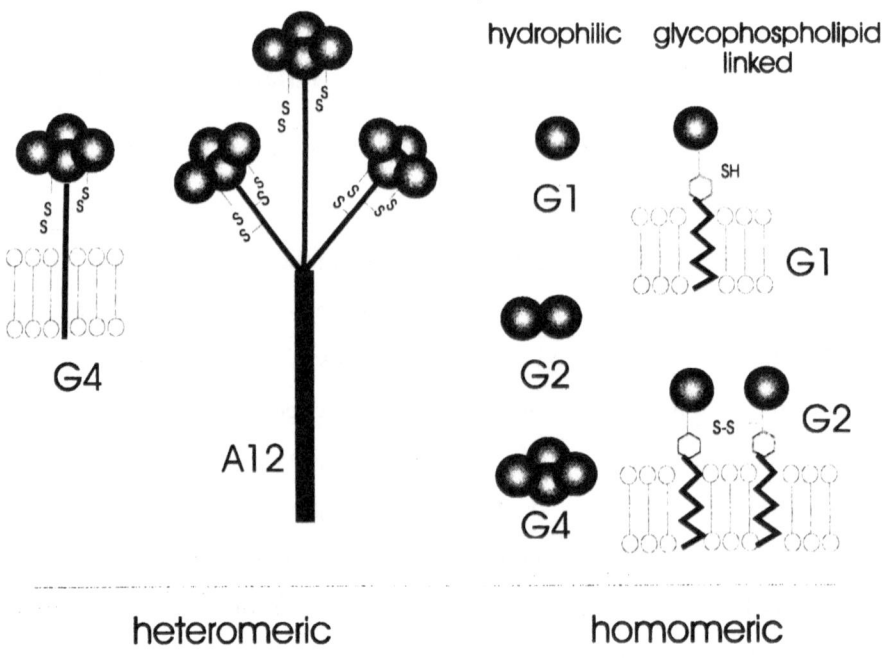

Fig. 5.1: Molecular forms of AChE (Massoulié and Toutant 1988)

As a result of its key function in organisms, AChE is an enzyme well-suited to screen for neurotoxins in water and food. As it can conveniently be isolated from several higher organisms, e.g. from the electric organ of the eel, and exhibits a good stability both during storage and under operation, numerous assay formats based on AChE inhibition have been developed for the detection of organophosphates and carbamates in food and water samples. In a typical assay, AChE is reacted for 30 min with a sample, and activity measurements at the beginning and at the end of the experiments serve as controls. For continuous assays, a fresh sample of AChE must be supplied after each measurement.

Most vertebrates express a second variety of cholinesterases with a different and less stringent specificity, usually termed butyrylcholinesterases (BuChE). As the biological function of these enzymes is not related to neurosignal transmission and generally less well understood, it will not be discussed here.

In the following chapters, we will focus on some biochemical properties of AChEs both from vertebrates and invertebrates, on their isolation in native or recombinant form, on their structure and mechanism as revealed by x-ray crystallography and protein engineering, and on some of the many assay formats described for AChE inhibition.

5.2   Biochemical properties of vertebrate AChEs

In vertebrates, AChE is present in multiple forms that are classified in two main categories: homomeric and heteromeric forms (Massoulié and Toutant 1988, Figure 5.1). The former are globular (G) and exist in two different varieties: (i) as an amphiphilic protein linked to the plasma membrane by phosphatidyl inositol glycerol and (ii) as a hydrophilic soluble protein which may occur as a monomer, dimer or tetramer. The heteromeric forms reveal more complex architectures: in muscle tissue, catalytic subunits associated in tetramers are linked to the basal lamina via a collagen tail (asymmetric forms , "A forms"), in nerve cells tetramers are linked to the membrane by a lipid ("G forms").

The catalytic subunit of all these forms is encoded by a single gene and the extensive polymorphism is due to alternative splicing of the mRNA and posttranslational modifications (Massoulié and Toutant 1988). The catalytic subunit typically comprises about 600 amino acids and contains seven cysteine residues, six of which form disulfides bridges important for the tertiary structure of the protein. Under ambient conditions, this structure is quite stable, and dried samples of snake venom were shown to exhibit some AChE activity even after several decades (Grassi, pers. com.). This surprising stability renders AChE a very good candidate for biotechnological applications, e. g. in the field of biosensors. Vertebrate AChEs exhibit Michaelis-Menten type kinetics and are inhibited by an excess of substrate (Radic et al. 1993, Szegletes et al. 1998).

5.3   Biochemistry of AChE from invertebrates

The molecular biology and the biochemistry of invertebrate AChEs have been less well studied than AChEs from vertebrates, and many aspects still remain unclear. The number of genes coding for AChEs and structural features of the protein may vary; one common feature to invertebrates, however, is the absence of A forms of the protein, which is common in vertebrates.

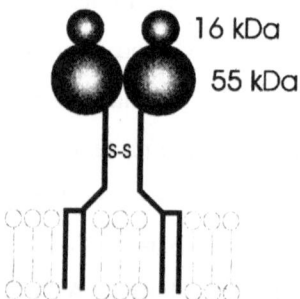

Fig. 5.2: Model for the *Drosophila* AChE structure (Fournier et al. 1988)

The major core of knowledge about insect AChEs has come from studies on *Drosophila melanogaster*. In this insect, there is just one gene encoding for synaptic AChE, and the gene product shows 31% homology with AChE from electric eel (Hall and Spierer 1986). *Drosophila* AChE is a G_2 amphiphilic dimer linked to the plasma membrane by a phospholipid (Fournier et al. 1988). Each subunit is formed by two peptides of 55 and 16 kDa linked in a non-covalent manner (Fournier et al. 1988, Figure 5.2).

The *Drosophila* enzyme can be stabilised using various procedures (Estrada-Mondaca 1998), facilitating its use for biotechnological purposes. Other than the vertebrate enzyme, *Drosophila* AChE does not follow a Michaelis-Menten type kinetic, but also displays substrate inhibition (Marcel et al. 1998, Stojan et al. 1998). It is more sensitive to insecticides than vertebrate AChE and thus is an interesting candidate for insecticide assays (Villatte et al. 1998). It has been possible, through site-directed mutagenesis in the active site, to increase its sensitivity (Villatte et al. 1998) and to create highly resistant variants (Villatte 1998).

Our knowledge about AChEs from other insects is more scattered. In the mosquito *Culex* (Bourguet et al. 1996), there are at least two different genes which encode AChEs exhibiting different sensitivity towards insecticides. Both genetic and biochemical evidence suggests the existence of two AChEs in the aphid *Aphis gossypii* (Villatte 1998). In the nematode *Caenorhabditis elegans*, there are four AChE genes (Grauso et al. 1998), and the gene products display different kinetic features (Johnson and Russel 1983). In the nematode *Nippostrongylus brasiliensis*, two different forms of AChE have been detected (Hussein et al. 1999). There are also at least two genes in *Amphioxus*, encoding two AChEs with different biochemical features, including their sensitivity towards inhibitors (McClellan et al. 1998). This wide diversity of AChEs provides an interesting reservoir for various pharmacological and biochemical applications which also might be used for analytical purposes.

5.4   Isolation, cloning and expression of AChE

Acetylcholinesterases are preferentially isolated from tissue which is easily available and in which the enzyme is abundant. From this practical point of view, preferred sources of the enzyme have been bovine erythrocytes and the electric organ of the eel, *Torpedo californica*. In the former case, the enzyme can be isolated from bovine blood, but purification to high specific acitivity depends on the removal of a C-terminal decapeptide by papain cleavage (Schmidt-Dannert et al. 1994). Using similar methods, AChE with a specific activity of several hundred units per mg protein can be obtained and indeed are commercially available.

A comfortable supply of AChEs from mammalian nervous tissue or from insects has only recently become available with the introduction of cloning techniques. In the past 10 years, a variety of recombinant AChEs were prepared from pertinent cDNA libraries. Table 5.1 provides some examples.

As can be seen from table 5.1, yields of recombinant AChEs are generally quite low. While this does not impede further investigations on the mechanism of this enzyme, it poses a severe limitation to practical applications, e. g. to the wider use of neurotoxin assays based on AChE inhibition. A major step forward might be the successful preparation of a synthetic gene encoding human brain AChE which has been expressed in *Pichia pastoris* (S. Vorlova et al., unpublished).

Table 5.1: Recombinant AChEs (examples)

Organism, tissue	Expression system	Expression level	Literature
Rat (brain)	COS cells	10^{-6} g/L	(Morel 1997)
	Baculovirus	10^{-6} g/L	(Mionetto et al. 1997)
	E. coli	10^{-6} g/mg inclusion bodies	(Heim et al. 1998)
	Pichia pastoris	10^{-3} g/L	(Heim et al. 1998)
Drosophila melanogaster	Baculovirus	10^{-3} g/L	(Chaabihi et al. 1994)

It should be added that the potential of enzymes for the analysis of neurotoxins is not necessarily limited to AChEs: in a recent study, Sigolaeva et al. describe a neuropathy target esterase (NTE) of unknown structure and sequence which forms in humans exposed to organophosphates and can be used in a sensitive biosensor assay for the detection of these compounds (Sigolaeva et al. 2000).

5.5 Structure and function of acetylcholinesterases

5.5.1 Sequence and structure

AChEs belong to a common homologous family, which has remote sequence similarity to other esterases, especially butyrylcholinesterases, carboxylesterases and lipases (Cousin et al. 1996, Pleiss et al.). DNA or mRNA of more than 50 AChEs have been sequenced, and 38 structure entries have been deposited in the PDB (Bernstein et al. 1977) since 1991, when the first experimental AChE structure became available (Sussman et al. 1991), including enzyme-inhibitor complexes, mutated AChEs, and homology models of AChE from *Torpedo californica* (28 entries), *Electrophorus electricus* (3 entries), *Drosophila melanogaster* (3 entries), *Mus musculus* (2 entries), and *Homo sapiens* (2 entries). Comprehensive information on sequence and structure of AChEs is provided by the well maintained ESTHER database (Cousin et al. 1996) at http://www.ensam.inra.fr/cholinesterase.

AChEs belong to the α/β hydrolase fold family (Ollis et al. 1992). The α/β hydrolase fold consists of a central hydrophobic eight-stranded β-sheet packed between two layers of amphiphilic α-helices. Serine esterases and lipases have a common catalytic mechanism of ester hydrolysis, which consists of five subsequent steps (Cygler et al. 1994): after binding of the ester substrate, a first tetrahedral intermediate is formed by nucleophilic attack of the catalytic serine, with the oxyanion stabilized by two or three hydrogen bonds, the so-called oxyanion hole. The ester bond is cleaved and the alcohol moiety leaves the enzyme. In a last step, the acyl enzyme is hydrolyzed. The nucleophilic attack by the catalytic serine is

112

mediated by the catalytic histidine and a third group (carboxylate groups of glutamic or aspartic acid side chain, or a carbonyl group of the protein backbone). Studies on the mechanism of AChE inhibition by organophosphates and carbamates have suggested that the enzyme is acylated by organophosphates and carbamates at a nucleophilic serine (Watts and Wilkinson 1977). The inhibition constant depends on the chemical structure of the substituents and the leaving group (Herzsprung et al. 1989). Other inhibitors bind non-covalently near the active site thus preventing a substrate molecule from binding.

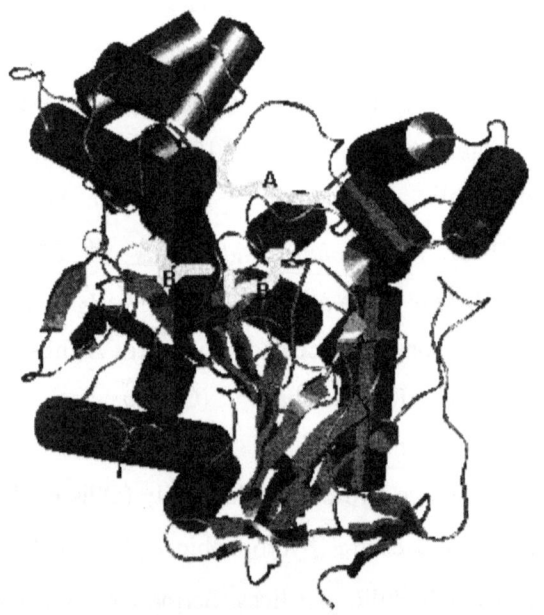

Fig. 5.3: Modelled structure of rat brain AChE; catalytic triad (B), acyl binding pocket (A)

5.5.2   Binding sites

Three residues in a "catalytic triad" are essential for the enzyme activity: a serine, a histidine and a glutamate (Rosenberry 1975, Duval et al. 1992). They are located in a 20 Å deep gorge forming the active site of the enzyme (Sussman et al. 1991).

This active site is located near the bottom of a deep and narrow gorge lined with 14 conserved aromatic amino acids (Saxena et al. 1997). Substrates and inhibitors bind to three distinct binding sites (Radic et al. 1993): (1) the size of the acyl pocket (F295, R296, F297, V300 in mouse AChE) determines substrate specificity (butyrylcholine $_{vs}$ acetylcholine) and mediates sensitivity toward transition state analog inhibitors organophosphate and carbamate of different size; (2) the choline binding site (W86, E202, Y337) binds specifically the substrate choline group and is blocked by tricyclic inhibitors (acridine, phenothiazine, and their derivatives); (3) the peripheral anionic site (Y72 , Y124, W286) it is located near the entrance to the substrate binding gorge and is blocked by charged, mono- or bisquaternary inhibitors (edrophonium, propidium, decamethonium).

Fig. 5.4: Complex of wild-type (left) and F295L mutant (right) rat brain AChE with paraoxon-ethyl. The catalytic triad (S 203, H 447, E 334) and the mutation site are displayed in gray and black, respectively

To investigate the relationship between sequence and specificity, the structures of complexes of AChEs with transition state analog inhibitors (Bartolucci et al. 1999, Millard et al. 1999, Silman et al. 1999) and inhibitors binding to the choline and the peripheral anionic binding site (Harel et al. 1993) have been determined. Comparing the structure of complexes with different inhibitors, it was observed that catalytic machinery and the side chains of the acyl binding pocket accommodate different inhibitors without major conformational changes (Silman et al. 1999). Based on structure data, the interactions between enzyme and inhibitors were further studied by computer-aided modelling and site-directed mutagenesis.

Molecular dynamics simulations indicated that access to the binding site is controlled by a gating mechanism: five aromatic rings rapidly reorientate which leads to rapid opening and closing of the gate (Zhou et al. 1998). Binding of inhibitors to the gorge is mediated by a hydrogen bonding network including water (Liu et al. 1998, Tara et al. 1999). The active site gorge was not rigid, but fluctuated in size and showed some plasticity: upon binding of bulky inhibitors, the diameter of the gorge increased slightly (Tara et al. 1999, Axelsen et al. 1994). The volume of the gorge is correlated with different sensitivities toward inhibitors. Since the gorge of butyrylcholinesterase (BuChE) is 200 $Å^3$ larger than that of AChE, inhibitors like ethopropazin have higher inhibitory activity toward BuChE (Saxena et al. 1999).

5.5.3   Acyl binding pocket

The role of the side chains of the acyl binding pocket was investigated by modelling and subsequent site directed mutagenesis. Increasing the acyl binding pocket of AChE by replacing F295 by tyrosine (Vellom et al. 1993) or leucine (Radic et al. 1993, Radic et al. 1994, Pleiss et al. 1997, Pleiss et al. 1999) had two effects: (1) The mutant hydrolyses butyrylcholine, while its catalytic activity toward acetylthiocholine decreased. (2) The mutant enzyme became more sensitive toward organophosphates; the ratio between k_i values of mutant and wild type increased with bulkiness of the inhibitor: 2, 7 and 80 for for paraoxon-methyl, paraoxon-ethyl and ethoprophos, respectively (Pleiss et al. 1999).

5.6   Assay formats

While routine analysis of organophosphate and carbamate insecticides in food and environmental samples is presently carried out by chromatographic methods, such as HPLC, GC, GC-MS or LC-MS/MS, these procedures are expensive, time-consuming and not suitable for field use (EPA Method 8141A, Lacorte and Barcelo 1995). In contrast, the detection and quantification of neurotoxic compounds using acetylcholinesterase is straightforward and well documented in the literature. All analytes which bind to and inhibit AChE can be measured by this method and comprise organophosphate and carbamate insecticides, nerve gases such as sarine and algal toxins such as anatoxin-a(s). Figure 5.5 shows a schematic view on the mechanism of cholinesterase inhibition by two major representants of organophosphates and carbamates. It is noteworthy, that the half-life period for spontaneous reactivation for diethoxyphosphorylated AChE from electric eel is 27 days in contrast to 38 min in the case of the methylcarbamylated enzyme (Main 1979). If the AChE is inhibited by organophosphates, an "ageing" phenomenon is observed: only for a limited time the phosphorylated enzyme can be reactivated by oximes such as 2-PAM. Thus, the possibility to reactivate AChE by 2-PAM after inhibition can be used as an argument for the presence of organophosphorus compounds, and is now often used in biosensor formats.

A simple test kit for the spectrophotometric assay of organophosphates and carbamates in water was described by Beutler (1993) and has been standardized in Germany in 1995 (DIN 38415-1). The test is based on the determination of AChE inhibition using a method first described by Ellman (Ellmann et al. 1961). It consists of the following steps:

1. Extraction of analytes using a liquid/solid extraction out of a water sample and evaporation of solvent (dichloromethane).

2. Oxidation of thiocompounds using n-bromsuccinimide and ascorbic acid as reducing agent.

3. Determination of AChE initial activity using acetylthiocholine as substrate and dithio-bis-nitrobenzoate as chromogen (Ellmann et al. 1961).

4. Incubation of AChE solution with the sample for 30 min.

5. Determination of residual AChE activity.

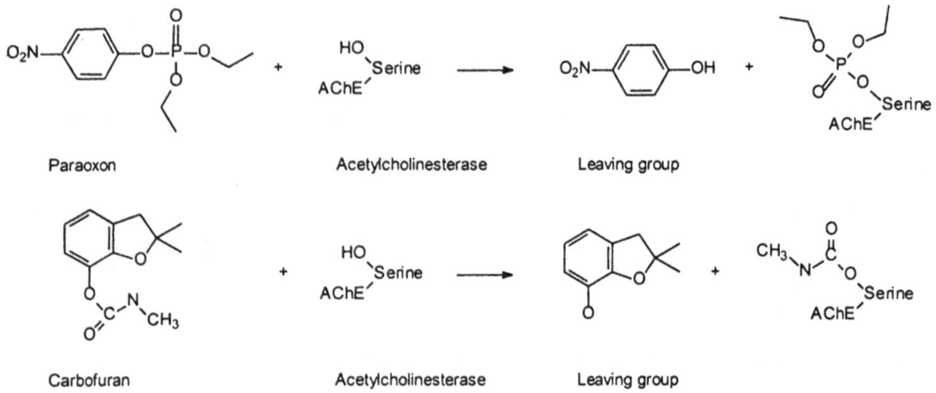

Fig. 5.5: Schematic view on AChE inhibition by paraoxon and carbofuran

AChE inhibition can then be calculated as percent inhibition $[\%] = ((a_0 - a_i) / a_0) * 100$. ($a_0$ = activity without inhibition, a_i = enzyme activity after incubation with sample) and is converted into paraoxon equivalent concentration by means of a standard curve. This simple spectrophotometric test is quite useful for water analysis but does not contain sample preparation steps prerequisite to food analysis. Other than this test kit format, a large number of biosensors employing immobilized cholinesterases were described. In addition to a commercial AChE-based sensor (Cide Lite from Charm Sciences, Malden, USA; Saul et al. 1995), numerous prototypes were developed, based on potentiometric (Ghindilis et al. 1996, Evtugyn et al. 1996), amperometric (Mionetto et al. 1994, Palleschi et al. 1992, Skladal and Mascini 1992) and piezoelectric (Guilbault and Ngeh-Ngwainbi 1988, Abad et al. 1998) transducers. Since enzymatic activity in AChE-based biosensors is destroyed upon calibration or in the presence of the analyte, the sensor must either be discarded after evaluation (Hartley and Hart 1994, Kulys and D'Costa 1991) or the enzyme must be replaced (Kindervater et al. 1990, Hart et al. 1997, Palchetti et al. 1997, Morelis and Coulet 1990). Usually, the amount of analyte is quantified via determination of AChE activity before and after incubation with the sample. The general test format is summarized in Figure 5.6.

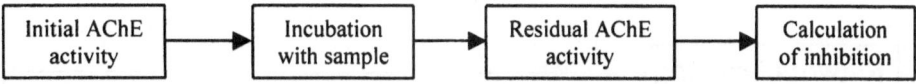

Fig. 5.6: Schematic view on neuroinhibitor detection by AChE inhibition tests

There are two major challenges to AChE-based inhibition analysis: specificity and accuracy.

Specificity: A severe drawback of all methods based on AChE inhibition assays stem from the fact that each organophosphate and carbamate inhibits this enzyme to a different extent, rendering calibration for an unknown mixture virtually impossible. As an example, the inhibition constants for individual organophosphates and carbamates against bovine erythrocyte AChE vary up to a factor of 500 (Herzsprung et al. 1989). In practice, calibration with paraoxon is generally used, but the value of this procedures for assays of AChE inhibitors of unknown composition raises doubts. On the other hand, all biosensors described so far are based on this mechanism, thus providing, as signal, a sum parameter expressed as total anti-acetylcholinesterase activity using paraoxon as reference. In order to escape this dilemma, a novel biosensor format capable to discriminate single cholinesterase inhibitors in a mixture was developed. It is based on disposable thick film electrodes and advanced data processing (Bachmann and Schmid 1999, Bachmann et al. 1999). Tests of binary mixtures of insecticides could be performed within one hour at high resolution. Either commercially available AChEs from natural sources or recombinant AChEs were used, all differing both in sensitivity and specificity. As such variants display a desirable cross-reactivity towards both analytes, the multisensor signals could be subjected to chemometric analysis, using artificial neural networks (ANN) (Seemann et al. 1997, Wittmann et al. 1997). Up to 4 different AChEs were immobilised by screen-

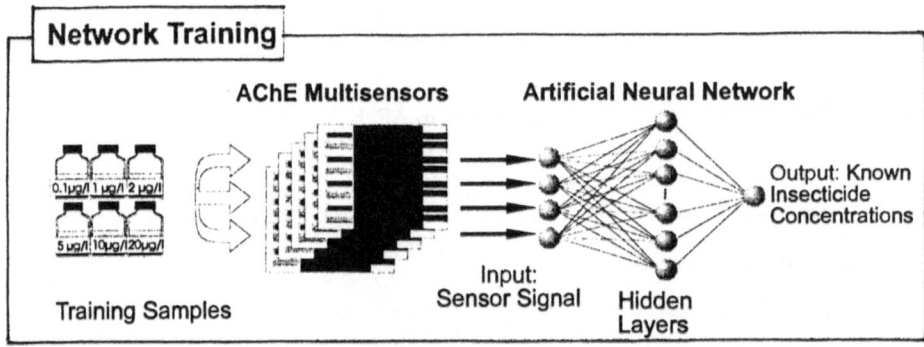

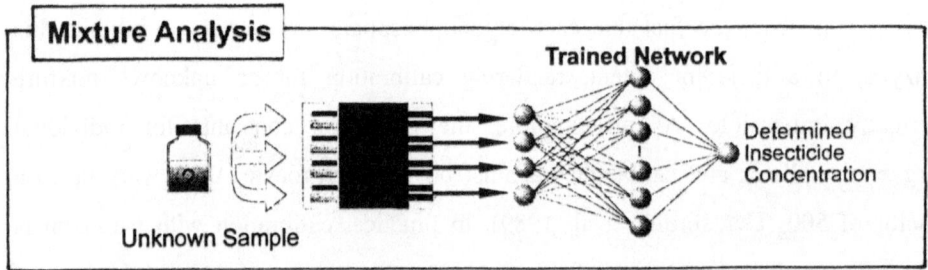

Fig. 5.7: Assay format for insecticide detection in mixtures using disposable AChE-multisensors

printing on four-electrode thickfilm sensors. Figure 5.7 shows the concept of this assay, starting with initial network training by measuring samples containing AChE inhibitors of known concentration ("network training") followed by the analysis of unknown samples ("mixture analysis").

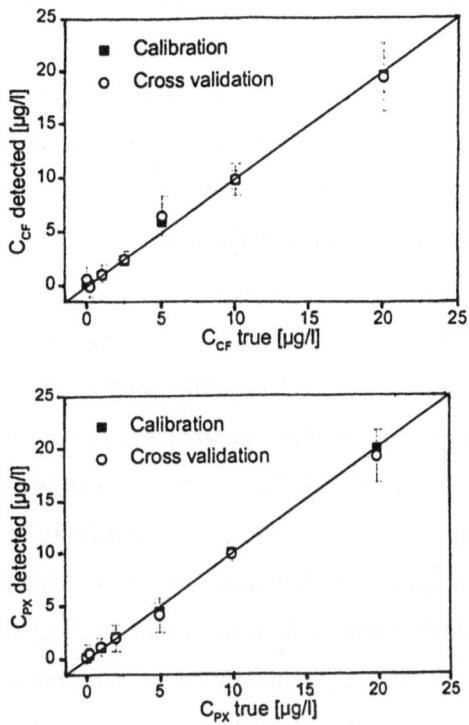

Fig. 5.8: Selective detection of carbofuran (CF) and paraoxon-ethyl (PX) in binary mixtures using the AChE-multisensor / artificial neural network system

With such type of multisensor, it was possible to discriminate quantitatively between paraoxon and carbofuran, two different classes of AChE inhibitors (Figure 5.8). The first version of this multisensor employed four types of native and recombinant AChEs (electric eel, bovine erythrocytes, rat brain, *Drosophila melanogaster*). The multisensor registered paraoxon and carbofuran in mixtures of 0 - 20 µg/l for each analyte with prediction errors of 0.9 µg/l for paraoxon and 1.4 µg/l for carbofuran within less than 60 min (Bachmann and Schmid 1999). Improved second generation AChE-multisensors were obtained with four different mutants of *Drosophila melanogaster* AChE (Bachmann et al. 1999), which were optimized by protein engineering for enhanced sensitivity (Villatte 1998).

Accuracy of detection: Another major drawback of AChE inhibition tests is that the analytical response is inversely related to the analyte concentration. Thus, for small inhibitor concentrations (TrinkwV BGBL 1986), one has to cope with the highest error. Makower et al. (1997) recently described the Affinity EnzymoMetric Assay (AEMA) which uses a chromatographic separation to eliminate this problem. The test principle is as follows (Figure 5.9): A bifunctional molecule („Bienzyme conjugate"), consisting of a highly specific acceptor (AChE) and a signal generating element (horseradish peroxidase) is mixed with the sample, containing organophosphates ("OP"). The mixture is then passed over an affinity surface, with ligands (paraoxon derivatives) of high affinity towards AChE. As a result, the bienzyme conjugate with bound analyte is separated from the free bienzyme conjugate, which binds to the affinity surface. The amount of bienzyme conjugate in the eluate is directly proportional to the amount of analyte in the sample and can be detected via horseradish peroxidase employing a chromogenic substrate (tetramethylbenzidine). Two assay formats, for steady-state and flow-injection measurements, were developed. In both cases, the lower detection limit for the model analyte diisopropylphosphate was as low as 1 pM. The level of unspecific signal did not exceed 12 % if no inhibitor was present in the sample. Furthermore, the number of steps necessary to perform an enzyme inhibiton test were reduced. This novel approach demonstrates the usefulness of an inversed detection mode for AChE inhibitors.

We consider the two concepts described above important steps to solve problems inherent to enzyme inhibiton analyis, namely accuracy and inhibitor discrimination.

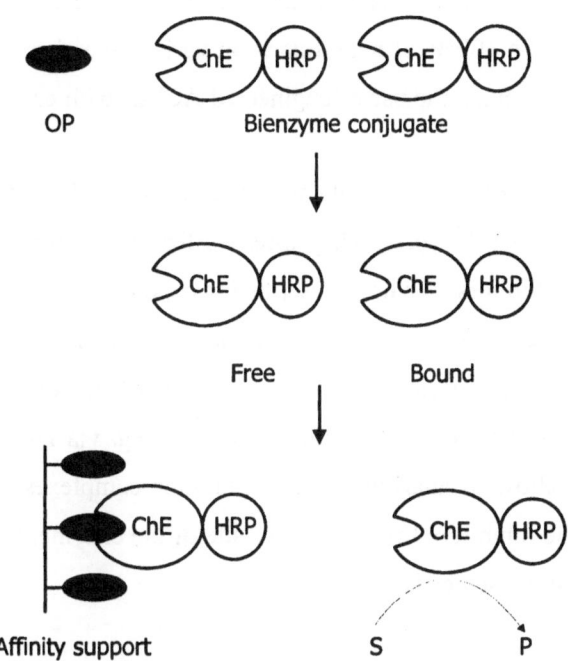

Fig. 5.9: Schematic principle of the Affinity EnzymoMetric Assay (AEMA). OP: organophosphate, ChE: cholinesterase, HRP: horseradish peroxidase, S: HRP substrate, P: product of HRP reaction (Makower et al. 1997)

5.7   Bioresponse-linked analysis based on AChE

Acetylcholinesterase is a promising element of biorecognition for the bioresponse-linked analysis of food and environmental samples. Due to its key role in neurotransmission, it can be considered a relevant target for the detection of AChE-specific neurotoxins such as organophosphates, carbamates and algal anatoxins. From the point of view of availability, AChEs of various origin can be prepared either by the extraction of animal tissue or in recombinant form through the expression of pertinent cDNA in suitable host organisms. While human brain AChE certainly deserves a prominent place as a target enzyme, AChEs of other origins, e.g. insect AChE, may add to the relevance of the assay, as they allow monitoring of putative effects of neurotoxins on various members of a food chain. This approach may include the use of engineered AChEs with enhanced specificity and selectivity.

A large number of assay formats and biosensors based on AChE have been described over the past 30 years. They often suffer from limits in accuracy and discrimination of inhibitors. We have described two examples which show progress in these areas. However, another challenge remains to be mastered before AChEs can be employed for the bioresponse-linked analysis of neurotoxins followed by the chemical analysis: formats of chemical analysis, e.g. via HPLC/MS, must be developed which allow to analyze AChE – inhibitor complexes as such or after displacement of the inhibitors from the active site of the enzyme. First attempts are underway to answer these needs.

5.8   References

Abad, J.M., Pariente, F., Hernández, L., Abruña, H.D., Lorenzo, E. (1998): Determination of organophosphorus and carbamate pesticides using a piezoelectric biosensor. Anal. Chem. 70, 2848-2855.

Axelsen, P.H., Harel, M., Silman, I., Sussman, J.L. (1994): Structure and dynamics of the active site gorge of acetylcholinesterase: synergistic use of molecular dynamics simulation and X-ray crystallography. Protein Sci. 3, 188-197.

Bachmann, T.T., Schmid, R.D. (1999): A disposable, multielectrode biosensor for rapid simultaneous detection of the insecticides paraoxon and carbofuran at high resolution. Anal. Chim. Acta 401, 95-103.

Bachmann, T.T., Schmid, R.D., Leca, B., Marty, J.-L., Vilatte, F., Fournier, D. (1999): Improved discrimination of organophosphate and carbamate insecticides using recombinant mutants of *Drosophila* acetylcholinesterase and a disposable multielectrode biosensor. Biosensors and Bioelectronics, in print.

Bartolucci, C., Perola, E., Cellai, L., Brufani, M., Lamba, D. (1999): "Back door" opening implied by the crystal structure of a carbamoylated acetylcholinesterase. Biochemistry 38, 5714-5719.

Bernstein, F.C., Koetzle, T.F., Williams, G.J., Meyer, E.F., Jr., Brice, M.D., Rodgers, J.R., Kennard, O., Shimanouchi, T., Tasumi, M. (1977): The protein data bank. A computer-based archival file for macromolecular structures. Eur. J. Biochem. 80, 319-324.

Beutler, H.O. (1993): pp 19-36. In: Biochemische Methoden zur Schadstofferfassung in Wasser (GDCh, F. W. i. d., ed.). VCH: Weinheim.

Bourguet, D., Raymond, M., Fournier, D., Malcolm, C.A., Toutant, J.P., Arpagaus, M. (1996): Existence of two acetylcholinesterases in the mosquito *Culex pipiens* (Diptera:Culicidae). J. Neurochem. 67, 2115-2123.

Chaabihi, H., Fournier, D., Fedon, Y., Bossy, J.P., Ravallec, M., Devauchelle, G., Cerutti, M. (1994): Biochemical characterization of *Drosophila melanogaster* acetylcholinesterase expressed by recombinant baculoviruses. Biochem. Biophys. Res. Commun. 203, 734-742.

Cousin, X., Hotelier, T., Lievin, P., Toutant, J.P., Chatonnet, A. (1996): A cholinesterase genes server (ESTHER): A database of cholinesterase-related sequences for multiple alignments, phylogenetic relationships, mutations and structural data retrieval. Nucleic Acids Res. 24, 132-136.

Cygler, M., Grochulski, P., Kazlauskas, R.J., Schrag, J.D., Bouthillier, F., Rubin, B., Serreqi, A.N., Gupta, A.K. (1994): A structural basis for the chiral preferences of lipases. J. Am. Chem. Soc. 116, 3180-3186.

DIN 38415-1 (1995): German standard methods for the examination of water, waste water and sludge.

Duval, N., Bon, S., Silman, I., Sussman, J., Massoulié, J. (1992): Site-directed mutagemesis of active-site-related residues in Torpedo acetylcholinesterase. Presence of a glutamic acid in the catalytic triad. FEBS Letts. 309, 421-423.

Ellmann, G.L., Courtey, K.D., Andres, V., Featherstone, R.M. (1961): A new and rapid determination of acetylcholinesterase activity. Biochem. Pharmacol. 7, 88-92.

EPA Method 8141A. In: EPA Method 8141A. US Environmental Protection Agency.

Estrada-Mondaca, S.a.F.D. (1998): Stabilisation of recombinant drosophila acetylcholinesterase. Prot. Exp. Purif. 12, 166-172.

Evtugyn, G.A., Budnikov, H.C., Nikolskaya, E.B. (1996): Influence of surface-active compounds on the response and sensitivity of cholinesterase biosensors for inhibitior determination. Analyst 121, 1911-1915.

Fournier, D., Bride, J.M., Karch, F., Bergé, J.B. (1988): Acetylcholinesterase form *Drosophila melanogaster*. Identification of two subunits encoded by the same gene. FEBS Letts. 238, 333-337.

Ghindilis, A.L., Morzunova, T.G., Barmin, A.V., Kurochkin, I.N. (1996): Potentiometric biosensors for cholinesterase inhibitor analysis based on mediatorless bioelectrocatalysis. Biosensors and Bioelectronics 11, 837-880.

Grauso, M., Culetto, E., Combes, D., Fedon, Y., Toutant, J.P., Arpagaus, M. (1998): Existence of four acetylcholinesterase genes in the nematodes *Caenorhabditis elegans* and *Caenorhabditis briggsae*. FEBS Letts. 424, 279-284.

Guilbault, G.G., Ngeh-Ngwainbi (1988): pp. 187. In: Analytical uses of immobilized biological compounds for detection, medical and industrial uses (Guilbault, G.G., Mascini, M., eds.). Reidel, The Netherlands, Dordrecht.

Hall, M.C., Spierer, P. (1986): The Ace locus of drosophila melanogaster: Structural gene for acetylcholinesterase with an unusual 5' leader. EMBO J. 5, 2949-2954.

Harel, M., Schalk, I., Ehret-Sabatier, L., Bouet, F., Goeldner, M., Hirth, C., Axelsen, P.H., Silman, I., Sussman, J.L. (1993): Quaternary ligand binding to aromatic residues in the active-site gorge of acetylcholinesterase. Proc. Natl. Acad. Sci. USA 90, 9031-9035.

Hart, A.L., Collier, W.A., Janssen, D. (1997): The response of screen printed enzyme electrodes containing cholinesterase to organophosphates in solution and from commercial formulations. Biosensors and Bioelectronics 12, 645-654.

Hartley, I.C., Hart, J.P. (1994): Amperometric measurement of organophosphate pesticides using a screen printed disposable sensor and biosensor based on cobalt phtalocyanine. Anal. Proc. Incl. Anal. Comm. 31, 333-336.

Heim, J., Schmidt-Dannert, C., Atomi, H., Schmid, R.D. (1998): Functional expression of a mammalian acetylcholinesterase in *Pichia pastoris*: Comparison to acetylcholinesterase, expressed and reconstituted from *Escherichia coli*. Biochim. Biophys. Acta 1396, 306-319.

Herzsprung, P., Weil, L., Quentin, K.E. (1989): Determination of organophosphorus compounds and carbamates by their inhibition of cholinesterase. Part 1: Inhibition values on immobilized cholinesterase. Zeitsch. Wasser u. Abwasser Forsch. 22, 67-72.

Hussein, A.S., Grigg, M.E., Selkirk, M.E. (1999): *Nippostrongylus brasiliensis*: Characterisation of a somatic amphiphilic acetylcholinesterase with properties distinct from the secreted enzymes. Exp. Parasitol. 91, 144-150.

Johnson, C.D., Russel, R.L. (1983): Multiple molecular forms of acetylcholinesterase in the nematode *Caenorhabditis elegans*. J. Neurochem. 41, 30-46.

Kindervater, R., Künnecke, W., Schmid, R.D. (1990): Exchangeable immobilized enzyme reactor for enzyme inhibition tests in flow-injection analysis using a magnetic device. Determination of pesticides in drinking water. Anal. Chim. Acta 234, 223-226.

Kulys, J., D'Costa, E.J. (1991): Printed amperometric sensor based on TCNQ and cholinesterase. Biosens. Bioelectr. 6, 109-115.

Lacorte, S., Barcelo, D. (1995): Determination of organophosphorus pesticides and their transformation products in river waters by automated on-line solid-phase extraction followed by thermospray liquid chromatography-mass spectrometry. J. Chrom. A5 712, 103-112

Liu, D.X., Jiang, H., Wang, Q.M., Chen, K.X., Ji, R.Y. (1998): Interpreting the effect of methyl group at the three carbon bridge of (-)-huperzine A on its anticholinesterase activity by molecular dynamics method. Bioorg. Med. Chem. Lett. 8, 419-422.

Main, A.R. (1979): Mode of action of acetylcholinesterases. Pharmacol. Ther. 6, 579-628.

Makower, A., Barmin, A., Morzunova, T., Eremenko, A.V., Kurochkin, I., Bier, F., Scheller, F.W. (1997): Affininty enzymometric assay for detection of organophosphorus compounds. Anal. Chim. Acta 357.

Marcel, V., Gagnoux Palacio, L., Pertuy, C., Masson, P., Fournier, D. (1998): Two invertebrate acetylcholinesteraes show activation followed by inhibition with substrate concentration. Biochem. J. 329, 329-334.

Massoulié, J., Toutant, J.P. (1988): Vertebrates cholinesterase: structure and types of interactions (Whittaker, V.P., ed.). Springer-Verlag. Berlin.

McClel.., J., Coblent, W.B., Sapp, M., Rulewicz, G., Gaines, D.I., Hawkins, A., Ozment, C., Bearden, A., Merritt, S., Cunningham, J., Palmer, E.,

Contractor, A., Pezzementi, L. (1998): cDNA cloning, in vitro expression, and biochemical characterization of cholinesterase 1 and cholinesterase 2 from amphioxus-comparison with cholinesterase 1 and cholinesterase 2, produced *in vivo*. Eur. J. Biochem. 258, 419-429.

Millard, C.B., Kryger, G., Ordentlich, A., Greenblatt, H.M., Harel, M., Raves, M.L., Segall, Y., Barak, D., Shafferman, A., Silman, I., Sussman, J.L. (1999): Crystal structures of aged phosphonylated acetylcholinesterase: nerve agent reaction products at the atomic level. Biochem. 38, 7032-7039.

Mionetto, N., Marty, J.-L., Karube, I. (1994): Acetycholinesterase in organic solvents for the detection of pesticides: Biosensor application. Biosens. Bioelectr. 9, 463-470.

Mionetto, N., Morel, N., Massoulié, J., Schmid, R.D. (1997): Biochemical determination of insecticides via cholinesterases. 1. Acetylcholinesterase from rat brain: functional expression using a baculovirus system, and biochemical characterization. Biotechnol. Tech. 11, 805-812.

Morel, N., Massoulie, J. (1997): Expression and processing of vertebrate acetylcholinesterase in the yeast Pichia pastoris. Biochem. J. 328, 121-129.

Morelis, R.M., Coulet, P.R. (1990): A sensitive biosensor for choline and acetylcholine involving fast immobilisation of a bienzyme system on a disposable membrane. Anal. Chim. Acta 231, 27-32.

Ollis, D.L., Cheah, E., Cygler, M., Dijkstra, B., Frolow, F., Franken, S.M., Harel, M., Remington, S.J., Silman, I., Schrag, J. (1992): The alpha/beta hydrolase fold. Protein Eng. 5, 197-211.

Palchetti, I., Cagnini, A., DelCarlo, M., Coppi, C., Mascini, M., Turner, A.P.F. (1997): Determination of anticholinesterase pesticides in real samples using a disposable biosensor. Anal. Chim. Acta 337, 315-321.

Palleschi, G., Bernabei, M., Cremisini, C., Mascini, M. (1992): Determination of organophosphorus insecticides with a choline electrochemical biosensor.

128

Sens. Act. B B7, 513-517.

Pleiss, J., Fischer, M., Peiker, M., Thiele, C., Schmid, R.D.: Lipase engineering database: Understanding and exploiting sequence-structure-function relationships. J. Mol. Catal. B: Enzym.

Pleiss, J., Mionetto, N., Schmid, R.D. (1999): Probing the acyl binding site of acetylcholinesterase by protein engineering. J. Mol. Catal. B6, 287-296.

Pleiss, J., Mionetto, N., Schmid, R.D. (1997): Protein engineering of rat brain acetylcholinesterase: a point mutation enhances sensitivity to pesticides. Prot. Eng. 10, 66-70.

Radic, Z., Pickering, N.A., Vellom, D.C., Camp, S., Taylor, P. (1993): Three distinct domains in the cholinesterase molecule confer selectivity for acetyl- and butyrylcholinesterase inhibitors. Biochem. 32, 12074-12084.

Radic, Z., Duran, R., Vellom, D.C., Li, Y., Cervenansky, C., Taylor, P. (1994): Site of fasciculin interaction with acetylcholinesterase. J. Biol. Chem. 269, 11233-11239.

Rosenberry, T.L. (1975): Acetylcholinesterase. Adv. Enzymol. Relat. Areas Mol. Biol. 43, 103-218.

Saul, A. J., Zomer, E., Puopolo, D., Charm, S.E. (1995): Use of new rapid bioluminescence method for screening organophosphate and n-methylcarbamate insecticides in processed baby foods. J. Food Proc. 59, 306-311.

Saxena, A., Redman, A.M., Jiang, X., Lockridge, O., Doctor, B.P. (1997): Differences in active site gorge dimensions of cholinesterases revealed by binding of inhibitors to human butyrylcholinesterase. Biochem. 36, 14642-14651.

Saxena, A., Redman, A.M., Jiang, X., Lockridge, O., Doctor, B.P. (1999): Differences in active-site gorge dimensions of cholinesterases revealed by binding of inhibitors to human butyrylcholinesterase. Chem. Biol. Interact.

119-120, 61-69.

Schmidt-Dannert, C., Kalisz, H.M., Safarik, I., Schmid, R.D. (1994): Improved properties of bovine erythrocyte acetylcholinesterase, isolated by papain cleavage. J. Biotechnol. 36, 231-237.

Seemann, J., Rapp, F.-R., Zell, A., Gauglitz, G. (1997): Classical and modern algorithms for the evaluation of data from sensor arrays. Fresenius J. Anal. Chem. 359, 100-106.

Sigolaeva, L.V, Makower, A., Eremenko, A.V., Makhaeva, G.F., Malygin, V.V., Kurochkin, I.N., Scheller, F. (2000): Bioelectrochmical analysis of neuropathy target esterase activity in blood. Bioanal. Chem. submitted.

Silman, I., Millard, C.B., Ordentlich, A., Greenblatt, H.M., Harel, M., Barak, D., Shafferman, A., Sussman, J.L. (1999): A preliminary comparison of structural models for catalytic intermediates of acetylcholinesterase. Chem. Biol. Interact. 119-120, 43-52.

Skladal, P., Mascini, M. (1992): Sensitive detection of pesticides using amperometric sensors based on cobalt phtalocyanin-modified composite electrodes and immobillized cholinesterases. Biosens. Bioelectr. 7, 335-343.

Stojan, J., Marcel, V., Estrada-Mondaca, S., Klaebe, A., Masson, P., Fournier, D. (1998): A putative kinetic model for substrate metabolisation by Drosophila acetylcholinesterase. FEBS Lett. 440, 85-8.

Sussman, J.L., Harel, M., Frolow, F., Oefner, C., Goldman, A., Toker, L., Silman, I. (1991): Atomic structure of acetylcholinesterase from Torpedo californica: a prototypic acetylcholine-binding protein. Science 253, 872-879.

Szegletes, T.M.W.D., Rosenberry, T.L. (1998): Nonequilibrium analysis alters the mechanistic interpretation of inhibition of acetylcholinesterase by peripheral site ligands. Biochem. 37, 4206-4216.

Tara, S., Helms, V., Straatsma, T.P., McCammon, J.A. (1999): Molecular

130

dynamics of mouse acetylcholinesterase complexed with huperzine A. Biopolymers 50, 347-359.

TrinkwV (1986): Verordnung über Trinkwasser und über Wasser für Lebensmittelbetriebe. BGBL I, 760-773.

Vellom, D.C., Radic, Z., Li, Y., Pickering, N.A., Camp, S., Taylor, P. (1993): Amino acid residues controlling acetylcholinesterase and butyrylcholinesterase specificity. Biochem. 32, 12-17.

Villatte, F. (1998): These de Doctorat, Université Paris 6, P. and M. Curie, Paris.

Villatte, F., Marcel, V., Estrada-Mondaca, S., Fournier, D. (1998): Engineering sensitive acetylcholinesterase for detection of organophosphate and carbamate insecticides. Biosens. Bioelectr. 13, 157-162.

Watts, P., Wilkinson, R.G. (1977): The interaction of carbamates with acetylcholinesterase. Biochem. Pharmacol. 26, 757-761.

Wittmann, C., Löffler, S., Zell, A., Schmid, R.D. (1997): pp. 343-352. In: Immunochemical Technology for Environmental Applications. American Chemical Society.

Zhou, H.X., Wlodek, S.T., McCammon, J.A. (1998): Conformation gating as a mechanism for enzyme specificity. Proc. Natl. Acad. Sci. USA 95, 9280-9283.

6 TRANSPORT PROTEINS IN BLOOD: A POSSIBLE ROLE IN HORMONE DISRUPTING EFFECTS OF POLLUTANTS

Eline P. Meulenberg

ELTI Support, Drieskensacker 12-10, 6546 MH Nijmegen, The Netherlands

Abstract. Hormone disrupting activities of pesticides and other enviromental contaminants are of great concern and many efforts have been made to develop assays for the assessment of such activities. Contaminants may affect hormonal systems at several levels. With regard to steroid and thyroid hormones it is known that there exists a delicate balance between the free and bound fraction in mammal blood, which defines the hormonal status of an organism. For this reason in the present contribution, attention is focussed on those proteins in blood that control the levels of bound and free steroid/thyroid hormones and hence there biological activities. Based on clinical studies relating to pathological conditions, the possible effects of pollutants through binding to blood proteins are explained.

Target proteins are the specific transport proteins SHBG (Sex Hormone Binding Globulin), CBG (Corticosteroid Binding Globulin) and TBG (Thyroxin Binding Globulin). With high affinity these proteins bind estrogens/androgens, corticosteroids/progesterone, and T3/T4, respectively. Based on these properties the possibilities to develop competitive assays using transport proteins as main component are discussed. Taking SHBG as an example, the results of a competitive binding assay for the assessment of the potential estrogenic/androgenic activity of various pesticides and pharmaceuticals are presented. Additionally, the effect of exogenous pollutants on metabolic, reproductive and psychological conditions in man as a result of displacement of endogenous hormones from their respective binding proteins are envisaged. Finally, future developments will be discussed, including application of binding assays in environmental and toxicological research and analysis.

6.1   Introduction

According to the definition hormone disruptors are compounds that interfere with endogenous hormone systems and as a result produce adverse effects. In mammals and especially humans there exist three major hormone systems that involve steroid and thyroid hormones (Hall et al. 1980). These hormones are produced in endocrine organs such as ovarium, testis, thyroid gland and adrenal; their main function is the regulation of metabolism, growth, differentiation, development, reproduction, homeostatis, behavior, neuronal maintenance, etc. (Hall et al. 1980, Wilson and Foster 1992, Clark et al. 1992). The mechanism of biological action of steroid and thyroid hormones has been described (Hall et al. 1980). In short, they are produced in the corresponding gland cells, secreted into the blood and transported to their respective target tissues/organs/cells. Here they bind to specific receptors that exert their function at the DNA level (Beato 1989). Exogenous compounds may influence the normal endocrine system at several levels: (1) at receptor level they may mimick the endogenous hormones or bind to the receptor and block the binding sites without having an effect; (2) at intracellular level they may inhibit or stimulate the synthesis of receptors; (3) at the level of the secreting glands they may inhibit or stimulate the synthesis of endogenous hormones; (4) at the level of hormone metabolism they may inhibit or stimulate those enzymes involved in hormone degradation; (5) in blood they may affect the transport of the hormones to their target cells (Zacharewski 1997, Soto et al. 1995, Jorgensen et al. 1998, Shelby et al. 1998, Siiteri 1982, Vermeulen 1986). In particular the point of transport proteins will be subject of this section (Pugeat et al. 1986).

6.2   Hormones in blood

Steroid and thyroid hormones are transported from their respective endocrine glands/organs to target cells through the blood. In blood they exist in a bound and free form. The greater part of the total amount of a specific hormone in blood is bound to proteins, whereas only a small but varying fraction is in the free state. According to the Free Hormone Hypothesis (Pardridge 1987, Mendel 1989, Ekins 1990, Robbins 1992), only this free fraction is biologically active, i.e. able to enter a target cell and exert its function. Although there are several reports indicating an additional function of protein-bound steroid and thyroid hormones (Rosner et al. 1986, 1999, Fortunati 1999), there are sufficient data to support the Free Hormone Hypothesis. Accordingly, the free fraction of hormones represents the most important part of the total concentration in blood. Any change in this free fraction, whether caused by a change in concentration of the particular hormone or the corresponding binding protein(s) will affect the hormonal status of a subject (Pugeat et al. 1981, 1987, Selby 1990). In the following the regulating mechanisms of the major hormonal systems will be shortly explained.

6.2.1   Corticosteroids

An important group of steroid hormones includes the corticosteroids with cortisol as main active compound. Cortisol is synthesized in the adrenal in response to ACTH (adrenocorticotrophic hormone) which in turn is released from the pituitary in response to CRH (corticotrophin releasing hormone). This is called the Hypothalamus-Pituitary-Adrenal-Axis (HPAA). The synthesis of cortisol occurs in a clear diurnal rhythm associated with a negative feedback mechanism. In blood from the total amount of cortisol the greater part is bound to mainly two proteins: to albumin, a protein with high capacity and low affinity, (6 – 10%), and to CBG, the specific transport protein with high affinity and low capacity (90 – 95%); the remaining portion (2 – 4%) is in the free state (Dunn et al. 1981). In normal situations the free concentration of cortisol is maintained between a defined lower and higher level (range 0.33 – 0.84 µg/L with a mean of 5.9 ± 1.7 µg/L (Meulenberg 1995). The role of CBG is to protect cortisol from degradation and to transport it to its target cells, where it can exert its function.

In women there are found situations wherein the levels of CBG and total cortisol are raised, whereas the level of free cortisol is in the normal range; these include pregnancy and the use of oral contraceptives (Meulenberg 1995). This can be explained by the increase in estrogens, either natural as in pregnancy or artificial from the oral contraceptives, which stimulate the synthesis of CBG in the liver, as is known from the literature (Sandberg and Slaunwhite 1959, Carol et al. 1980). An increase in CBG will lower the free fraction of cortisol, leading to stimulation of the release of ACTH and, consequently, the biosynthesis of cortisol in the adrenal and in turn more cortisol bound to CBG. In certain diseases, e.g. Cushing syndrome or depression, the level of CBG is in the normal range whereas total cortisol and concomittantly free cortisol are raised. In such situations there are observed obvious symptoms of hypercortisolism. In general, glucocorticoids are involved in the function of the immune system, metabolism, behavior, homeostasis. Increased cortisol levels have been related to obesitas, insuline resistance, HDL cholesterol and cardiovascular risk (Hall et al. 1980, Fraser et al. 1999).

It should be noted that in addition to cortisol, CBG also functions as high affinity binding protein for progesterone, a steroid hormone involved in reproduction. Progesterone is synthetized in the adrenal and ovaries in response to gonadotrophins. It is regulated in a negative (male) or combined negative and positive (female) feedback system at hypothalamic and pituitary level. In women the variation in progesterone is subject to the menstrual cycle and it plays an important role in pregnancy, during which period its concentration steadily increases until term. CBG has only one binding site and both cortisol and progesterone are bound to this single site, the affinity of cortisol being about three times that of progesterone (Dunn et al. 1981).

6.2.2   Androgens/Estrogens

Reproduction, sexual differentiation and behavior in mammals depends on levels and ratios of androgens and estrogens. These steroid hormones are synthesized in testes, ovaries and adrenals. Taking testosterone as an example, the mechanism of regulation again involves the hypothalamus that releases LHRH (Luteinising Hormone Releasing Hormone), which stimulates the pituitary to release LH (Luteinising Hormone) which in turn is the stimulus for the testicular synthesis of testosterone (Hall et al. 1980). This is defined as the Hypothalamus-Pituitary-Testis-Axis (HPTeA), a negative feedback system. Additionally, testosterone is periferally converted into estradiol in (fat) tissues. Testosterone is synthesized in a diurnal rhythm. In view of the importance of the HPTeA, the regulatory mechanism of the biosynthesis of testosterone is illustrated in Figure 6.1. In blood testosterone is distributed comparable to cortisol. The total concentration of testosterone includes the fraction bound with low affinity to albumin (30 – 50%), that bound with high affinity to SHBG (Sex Hormone Binding Globulin, 44 – 66%), and the unbound fraction (1 – 2%) (Dunn et al. 1981). The main function of SHBG is to protect the hormones against degradation and to transport them to their respective target cells. SHBG is a dimeric protein composed of two monomeric subunits containing one steroid binding site inbetween (Petra 1991). It is synthesized in the liver and its production is stimulated by estrogens (Johnson 1984, Joseph 1994, Kouretas et al. 1999). The affinity of testosterone for SHBG is about 3 times that of estradiol (Anderson 1974, Dunn et al. 1981). Abnormal levels of testosterone are found in several disorders. For example, hypogonadal men exhibit decreased concentrations and virilism in women is associated with increased concentrations (Anderson 1974, Siiteri et al. 1982, Johnson 1984).

136

Furthermore, breast cancer and testicular/prostatic cancer are related to estradiol and testosterone, respectively (Englebienne 1989).

HPTeA = Hypothalamus-Pituitary-Testis-Axis

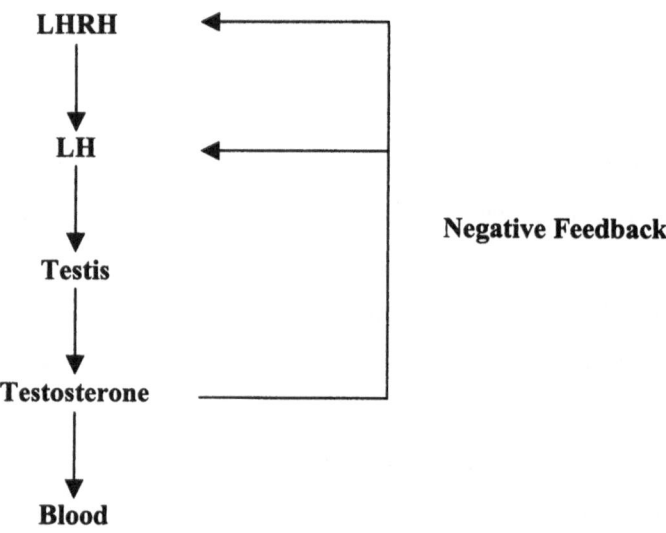

Fig. 6.1: Regulation of the biosynthesis of testostosterone in the Hypothalamus-Pituitary-Testis-Axis; LHRH = Luteinizing Hormone Releasing Hormone, released from the hypothalamus; LH = Luteinizing Hormone, released from the pituitary

6.2.3   Thyroids

The last group of hormones includes the thyroid hormones, particularly T3 (tri-iodothyronine) and T4 (tetra-iodothyronine or thyroxine). The production of these hormones is regulated by the Hypothalamus-Pituitary-Thyroid-Axis (HPTyA) in a negative feedback fashion and subjected to a diurnal rhythm (Hall et al. 1980). In this case, TRH (Thyrotrophin Releasing Hormone) is released from the hypothalamus, it stimulates the release of TSH (Thyroid Stimulating Hormone), which in turn activates the thyroid gland to synthesize T3 and T4. Additionally, periferally T4 is converted into T3. Once secreted into blood both hormones are for the greater part bound to proteins, i.e. low affinity binding albumine (12 – 18%), high affinity binding TBG (Thyroxine Binding Globulin, 73 – 82%) and an additional protein designated TBPA (Thyroxine Binding Pre-Albumin, 6 – 9%) (Hall et al. 1980, Mendel 1989). This latter protein is predominantly involved in the transport of thyroid hormones through the blood-brain barrier. Only a very small fraction, 0.018 – 0.024%, is in the free state in blood. TBG has the same function as CBG and SHBG; it protects T3/T4 against degradation and transports these hormones to their target cells. TBG contains only one binding site for both T3 and T4, and the affinity of T4 is about 50 times higher than of T3 (Hall et al. 1980). In normal conditions thyroids regulate basal metabolism, growth, development, neuronal maintenance, behavior. Several disorders are related to abnormal thyroid concentrations; for example, in pregnancy T4 is increased but without major symptoms. In thyroid cancer, Hashimoto's disease T4 and in stress T3 are clearly increased, whereas obesitas and mental retardation are accompanied with decreased levels of T4 (Hall et al. 1980, Divi and Doerge 1996, Sher et al. 1998).

6.3   Effects of hormone disrupting xenobiotics

Xenobiotic substances may enter the living body by ingestion, inhalation or transdermally. If a particular compound enters the blood compartment it may exhibit hormone-like properties with respect to the specific hormone-binding globulins CBG, SHBG and/or TBG. It may mimick one or more of the endogenous hormones and compete for the binding site at the respective binding globuline or it may alter the protein configuration such that any bound endogenous hormone is released. In both situations, the free concentration of the displaced or otherwise released hormone is increased. This may lead to a corresponding biological effect as mentioned above. The concept of inhibition of binding of endogenous hormones to SHBG by environmental xenobiotics has been described and investigated by Danzo (1997, 1998). Using post-partum serum as a source of SHBG and [^{3}H]-5-alpha-DHT as labeled ligand the effect of several xenobiotics on binding of the ligand was determined. It was demonstrated that hexachlorocyclohexane, o,p'-DDT, pentachlorophenol and nonylphenol exhibited binding to SHBG. In contrast, no such activity was found for metoxychlor, atrazine, p,p'-DDT, p,p'-DDE and dieldrin. It was also postulated that additional to displacement of endogenous hormones from SHBG, binding of pesticides could reduce the availability of these compounds for receptor binding. This may be especially relevant for pesticides which act through the estrogen or androgen receptor. Another group of pesticides, the pyrethroids, was studied by Eil and Nisula (1990). Pyrethroids can lead to gynecomastia in men due to their anti-androgenic action. They bind to the androgen receptor competitively and some of them, such as pyrethrins and bioallethrin, displace testosterone from SHBG. Further, the very toxic pollutant dioxin is reported to have a hormone deregulating effect by stimulating the conversion of testosterone into estradiol, leading to an increase in SHBG and concomittantly to raised insulin levels (Michalek et al. 1999). Not only xenobiotics may exert an estrogenic activity at the level of SBHG. It has also been reported that phytoestrogens can act like natural estrogens at several levels: binding to the estrogen receptor, induction of SHBG synthesis, stimulation of MCF-7 cells (Adlercreutz et al. 1987, Kurzer and Xu 1997, Adlercreutz 1999).

With regard to the thyroid hormones, PCB's, polyhalogenated aromatic hydrocarbon (PHAH) congeners and dioxin are indicated as neurotoxicants due to their thyroid hormone antagonistic or agonistic effects (Leatherland 1998, Sher et al. 1998, Porterfield and Hendry 1998). PCB's show a structural resemblance to T3/T4 and they bind to the thyroid receptor and TBG resulting in displacement of endogenous thyroid hormones leading to an enhanced metabolic clearance.

Additionally, they exert an adverse effect on thyroid-related enzymes. PHAH have been implicated in thyroid problems in fish-eating human populations in regions around the Great Lakes. These substances were mentioned to have affinity for TBG and an effect on the enzyme that converts T4 into T3. An anti-thyroid effect has also been reported for flavonoids which seem to inhibit thyroid peroxidase and as a result thyroid hormone synthesis (Divi and Doerge 1996, Divi et al. 1997). Unfortunately, no reports about binding of phytoestrogens to TBG are known.

Taking substances with an estrogen-like structure as an example, such compounds may displace estradiol from SHBG, thereby increasing the concentration of free estradiol. Any substance with a testosterone-like structure will have the same effect, due to the higher affinity of testosterone compared to estradiol for SHBG. Consequently, more estradiol will reach its target cells resulting in an enhanced effect. It will be appreciated that with regard to the other binding proteins the same mechanism may work. Corticosteroid-like compounds may displace cortisol from CBG and thyroid-like compounds may displace T3 and/or T4 from TBG.

6.4   Assays for hormone disrupting activity

6.4.1   SHBG assay

In the scope of potential hormone disrupting activities of xenobiotics and the need for assays to assess such activities, in particular estrogenic activities, an SHBG assay was designed based on the fact that this transport protein defines androgenic status in men. In the following this assay is decribed and preliminary results are shown. It is known that SHBG levels increase in the course of pregnancy and for that reason third trimester pregnancy serum was used as a source of this binding globulin. Using SHBG as high affinity binding component the assay was developed

140

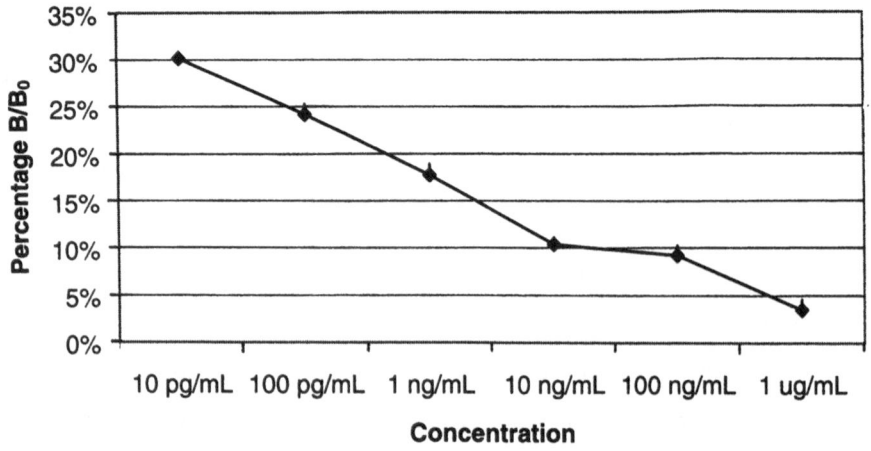

Fig. 6.2: Standard curve of testosterone. Microtiter plates coated with anti-SHBG antibody and SHBG were incubated with trititated estradiol (± 6000 cpm) and increasing concentrations of testosterone; the supernatant was measured and B/B_0 (bound cpm over bound at zero concentration) calculated; values are given as the mean of 11 runs ± % C.V

similar to an immunoassay on microtiter plate with labeled estradiol, $[1,2\text{-}^3H]\text{-}17\beta$-estradiol, as ligand. Because serum contains several other proteins that may bind estradiol, first the plate was coated with polyclonal anti-SHBG antibody. After incubation with serum and washing of the plate, labeled estradiol and sample were added and incubated during 48 hours. Then the free fraction of labeled estradiol was measured and the bound fraction of sample was calculated. For comparison of the activity of substances to be measured with the known activity of endogenous steroid hormones testosterone was taken as standard. Testosterone showed a clear displacement of labeled estradiol. In Figure 6.2 the mean of 11 standard curves is presented and the detection limit at 90% binding (expressed as B/B_0) of testosterone was 10 pg/mL.

Similarly, several pesticides and pharmaceuticals were tested in the assay. Figure 6.3 shows the results for a number of pesticides and Figure 6.4 for some pharmaceuticals.

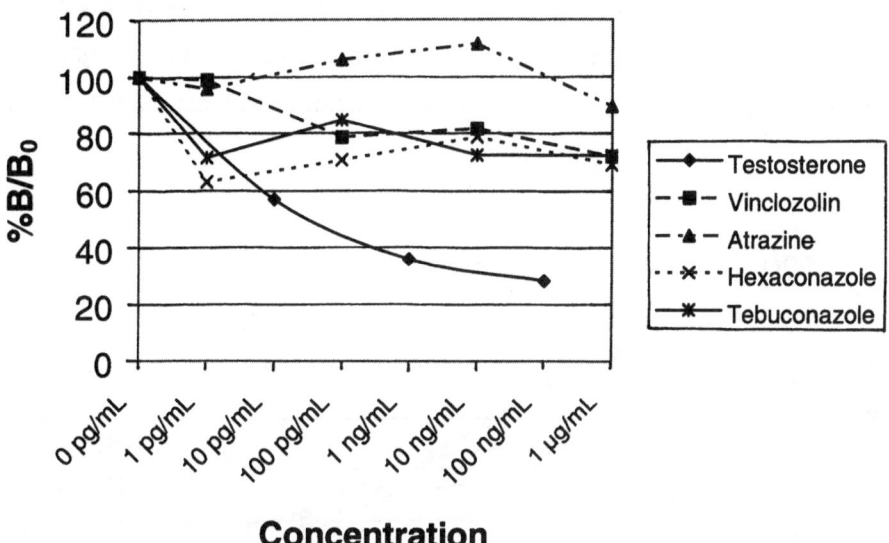

Concentration

Fig. 6.3: Assay of pesticides. Microtiter plates coated with anti-SHBG antibody and SHBG were incubated with tritiated E2 and increasing concentrations of test compounds; the supernatant was counted and bound fraction was calculated on the basis of total counts added

Among all the pesticides tested, it appeared that permethrin exhibited a pronounced estrogenic effect; malathion, 2,4-D, dichlofluanid and vinclozolin also had a positive effect, although to a lesser extent, whereas no displacement of estradiol was found for atrazine. The results for hexaconazole, tebuconazole, glyphosate and simazin were ambiguous. The displacement of tritiated estradiol by permithrin is in contrast to the findings of Eil and Nisula (1990); however, this may be explained by the fact that these workers used testosterone as ligand, which has a higher affinity for SHBG and thus requires higher concentrations of competitor than estradiol. An estrogenic effect of vinclozolin has been reported by Danzo (1998) and according to the findings of Soto et al. (1995) 2,4-D shows an anti-androgenic effect in the E-screen assay. Up to now, no data for glyphosate, dichlofluanid and simazin are available; the lack of displacement by atrazine is in agreement with the results of Danzo (1997). For aldicarb also no data are available, but an effect might be expected on the basis of adverse effects of carbamates on human pregnancy as reported by Savitz et al. (1997). Conazole pesticides have not been tested for

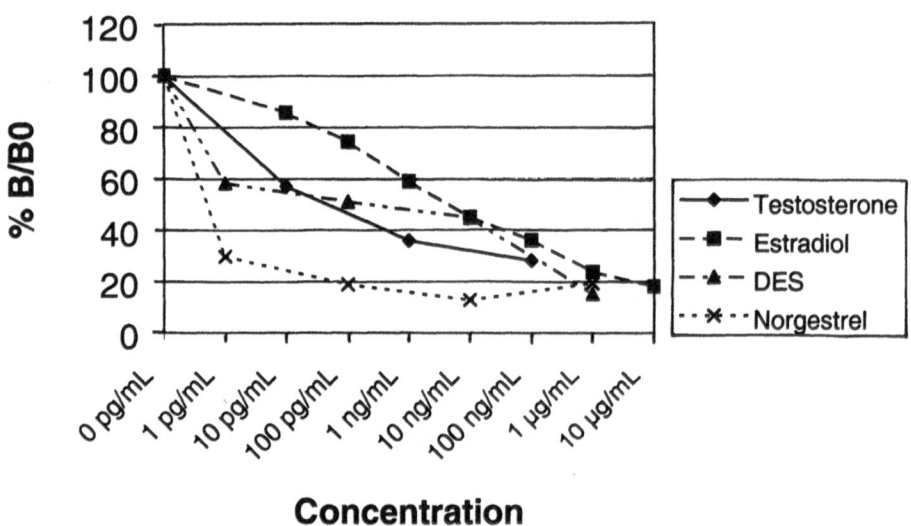

Fig. 6.4: Assay of pharmaceuticals. For details see Figure 6.3

SHBG binding before, but in clinical pharmacology various conazoles, in particular ketoconazole, are used as medicine in the treatment of disorders related to hyperandrogenism (Vidal-Puig et al. 1994). With regard to the pharmaceuticals tested, estradiol, DES and norgestrel showed high affinity for SHBG, while dehydro-iso-androsterone, androsterone and 6-alpha-methyl-prednisolone were able to displace the tracer to a lesser degree. The binding found for norgestrel is in accordance with the data as reported by Pugeat et al. (1981). Displacement by DES is unexpected, because the action of this compound is presumed to occur through the estrogen receptor or estradiol binding protein, which appears to be present in pregnancy serum (Englebienne 1986). This protein is not expected to interfere in our assay due to the use of anti-SHBG antiserum. For the other pharmaceutical compounds no data are available. Additionally, bis-phenol A and nonylphenol as industrial pollutants were used in the assay. Both substances showed inhibition of estradiol binding which is in agreement with literature data about an estrogenic effect of alkylphenols (Toppari et al. 1996, Daston et al. 1997, Danzo 1998), but in very high concentrations. Similar results for displacement of testosterone and estradiol from SHBG by bisphenol A and nonylphenol were reported by Déchaud et al. (1998) who used Concanavalin A-Sepharose as binding agent for SHBG from serum. A problem encountered with these compounds was their solubility in aqueous solution. To be able to make serial dilutions 5% ethanol was added, which appeared to have an effect on the standard curve of testosterone.

6.4.2   Application of the SHBG assay in environmental analysis

From the above results it is clear that the SHBG assay is able to assess a potential estrogenic activity of essentially any compound of interest. For example, in the scope of the regulatory admission of new pesticides or the re-examination of already admitted pesticides the determination of a potential estrogenic activity is recommended to be included as part of the toxicity test regimen. The SHBG assay may thus be used additionally to the estrogen receptor assay to obtain data at two levels of hormone disrupting activity. The activity of known compounds can be readily determined by applying them directly in the assay in various concentrations. Additionally, it may also be interesting to investigate various surface, ground or wastewater samples for estrogenic properties. If a positive result is found, it has to be determined which of the substances present is (are) the causitive component(s). To solve this question any positive sample can be further analyzed by HPLC or LC-MS and subsequently the various fractions can be applied again in the SHBG assay separately. As an alternative, water samples may be separated first by HPLC and the estrogenic activity of the eluted fractions determined subsequently. In view of the thousands of compounds present in surface water, which all may show more or less estrogenic activity, it is desirable to limit the number of components/fractions to be measured. This may be accomplished by introducing a prepurification step, for example, SPE extraction or (immuno)affinity chromatography before HPLC analysis. Even an estrogen receptor column may be used to select only those compounds that show affinity for this receptor.

6.4.3   Alternative assays for hormone disrupting compounds

The SHBG assay described above is suitable for the assessment of a potential estrogenic activity of known and unknown samples. However, pollutants may also exhibit an effect on other hormone systems such as the HPAA or HPTyA. To assess a potential corticosteroidal or thyroidal activity the corresponding assay may be designed comparable to the SHBG assay. Herein CBG or TBG may be used as the high affinity binding component and labeled cortisol or T3/T4, respectively, as ligand to determine displacement by substances or samples of interest. Any corticoid or thyroid action of xenobiotics may be implicated in observed phenomena in the human population of increase incidence of obsesitas, reproductive failure, diabetes, shortness of memory, stress symptoms, cadiovasulcar risk etc.

6.5   Conclusion

Great concern about hormone disrupting activities of xenobiotics has been expressed and there exists a need for assays giving a measure for any such activity. Several assays already exist and are being used, in particular those employing the estrogen receptor. Based on the knowledgde in the field of clinical chemistry, endocrinology and pharmacology, it can be concluded it is also important to determine an effect at the level of hormone binding proteins. Indications herefore have already been given by Danzo (1997, 1998) and Adlercreutz (1987, 1999). In this respect the SHBG assay decribed above appears to be a useful tool for assessing an estrogenic effect of environmental pollutants. The preliminary results shown are in good agreement with previous data and this supports the applicability of the assay to screen for a potential estrogenic activity of known and unknown substances. The SHBG assay should be regarded as supplemental to already existing assays, giving information about effects at a different level of biological activity. It has been mentioned that besides the HPTeA, major hormonal systems include the HPAA and HPTyA. Therefor, it may be desirable to develop similar assays based on CBG and TBG.

6.6   Future investigations

The applicability of the SHBG assay for the assessment of a potential estrogenic activity of xenobiotics has been demonstrated. However, there have been tested only a limited number of compounds. In further experiments attention will be focussed on additional pesticides with reported and unreported estrogenic or androgenic activity. Furthermore, synergistic effects should be determined by testing several combinations of pesticides. Other interesting groups of substances to be analyzed are presented by the phytoestrogens, industrial pollutants such as alkylphenols, PCB etc., and pharmaceuticals and drugs that are used in large amount, e.g. parent compounds and metabolites derived from oral contraceptives. Finally, in order to be used for complex environmental samples, such as surface or waste water, the SHBG assay should be coupled to conventional analytical methods including separation of components present in a positive sample and optionally to characterize unknown components. In this respect an off-line or on-line coupling with HPLC or LC-MS may be considered.

6.7 References

Adlercreutz, H., Höckerstedt, K., Bannwart, C., Bloigu, S., Hämäläinen, E., Fotsis, T., Ollus, A. (1987): Effect of dietary components, including lignans and phytoestrogens, on enterohepatic circulation and liver metabolism of estrogens and on sex hormone binding globulin (SHBG). J. Steroid Biochem. 27, 1135-1144.

Adlercreutz, H. (1999): Phytoestrogens. State of the art. Environ. Toxicol. Pharmacol. 7, 201-207.

Anderson, D.C. (1974): Sex-hormone-binding globulin. Clin. Endocrinol. 3, 69-96.

Beato, M. (1989): Gene regulation by steroid hormones. Cell 56, 335-344.

Carol, W., Börner A., Klinger, G., Greinke, C. (1980): Transcortin as an indicator of estrogenic potency in oral contraceptives. Endocrinology 75, 167-172.

Clark, J.H., Schrader, W.T., O'Malley, B.W. (1992): Mechanisms of action of steroid hormones, p. 35-90. In: Williams Textbook of Endocrinology (Wilson, J.D., Foster, D.W., eds.). W.B. Saunders Comp., Philadelphia.

Danzo, B.J. (1997): Environmental xenobiotics may disrupt normal endocrine function by interfering with the binding of physiological ligands to steroid receptors and binding proteins. Environ. Health Perspect. 105, 294-301.

Danzo, B.J. (1998): The effects of environmental hormones on reproduction. Cell. Molec. Life Sci. 54, 1249-1264.

Daston, G.P., Gooch, J.W., Breslin, W.J., Shuey, D.L., Nikiforov, A.I., Fico, T.A., Gorsuch, J.M. (1997): Environmental estrogens and reproductive health: A discussion of the human and environmental data. Reprod. Toxicol. 11, 465-481.

Déchaud, H., Ravard, C., Claustrat, F., Brac de la Perrière, A., Pugeat, M. (1998): Xenoestrogen interaction with human sex hormone-binding globulin (hSHBG). Steroids 64, 328-334.

Divi, R.L., Doerge, D.R. (1996): Inhibition of thyroid peroxidase by dietary flavonoids. Chem. Res. Toxicol. 9, 16-23.

Divi, R.L., Chang, H.C., Doerge, D.R. (1997): Anti-thyroid isoflavones from soybean. Biochem. Pharmacol. 54, 1087-1096.

Dunn, J.F., Nisula, B.C., Rodbard, D. (1981): Transport of steroid hormones: binding of 21 endogenous steroids to both testosterone-binding globulin and corticosteroid-binding globulin in human plasma. J. Clin. Endocrinol. Metab. 53, 58-68.

Eil, C., Nisula, B. (1990): The binding properties of pyrethroids to human skin fibroblast androgen receptors and to sex hormone binding globulin. J. Steroid Biochem. 35, 409-414.

Ekins, R. (1990): Measurement of free hormones in blood. Endocrine Rev. 11, 5-46.

Englebienne, P. (1986): Diethylstilbestrol binding in pregnancy plasma. Clin. Chem. 32, 574-575.

Englebienne, P. (1989): Human serum steroid-binding proteins and malignancy. Anticancer Res. 9, 1769-1776.

Fortunati, N. (1999): Sex hormone-binding globulin: Not only a transport protein. What news is around the corner? J. Endocrinol. Invest. 22, 223-234.

Fraser, R., Ingram, M.C., Anderson, N.H., Morrison, C., Davies, E., Connell, J.M.C. (1999): Cortisol effects on body mass, blood pressure, and cholesterol in the general population. Hypertension 33, 1364-1368.

Hall, R., Anderson, J., Smart, G.A., Besser, M. (1980): Fundamentals of clinical endocrinology. Pitman Medical Ltd., Tunbridge, Kent.

Johnson, P. (1984): Sex hormones and the liver. Clin. Sci. 66, 369-376.

Jorgenssen, M., Hummel, R., Bévort, M., Andersson, A-M., Skakkebaek, N.W., Leffers, H. (1998): Detection of oestrogenic chemicals by assaying the expression level of oestrogen regulated genes. APMIS 106, 245-251.

Joseph, D.R. (1994): Structure, function, and regulation of androgen-binding protein/sex hormone-binding globulin. Vitamines and Hormones 49, 197-280.

Kouretas, D., Laliotis, V., Taitzoglou, I., Georgellis, A., Tsantarliotou, M., Mougios, V., Amiridis, G., Antonoglou, O. (1999): Sex-hormone binding globulin from sheep serum: Purification and effects of pregnancy and treatment with exogenous estradiol. Compar. Biochem. Physiol. 123 Part C, 233-239.

Kurzer, M.S., Xu, X. (1997): Dietary phytoestrogens. Annu. Rev. Nutr. 17, 353-381.

Leatherland, J.F. (1998): Changes in thyroid hormone economy following consumption of environmentally contaminated Great Lakes fish. Toxicol. Indu. Health 14, 41-57.

Mendel, C.M. (1989): The free hormone hypothesis: A physiologically based mathematical model. Endocrine Rev. 10, 232-274.

Meulenberg, E.P. (1995): Corticosteroids in plasma and saliva: The influence of oral contraceptive use and pregnancy. Thesis, University of Wageningen, The Netherlands.

Michalek, J.E., Akhtar, F.Z., Kiel, J.L. (1999): Serum dioxin, insulin, fasting glucose, and sex hormone-binding globulin in veterans of operation ranch hand. J. Clin. Endocrinol. Metab. 84, 1540-1543.

Pardridge, W.M. (1987): Plasma protein-mediated transport of steroid and thyroid hormones. Am. J. Physiol. 252 P1, E157-E164.

Petra, P.H. (1991): The plasma sex steroid binding protein (SBP or SHBG). A critical review of recent developments on the structure, molecular biology and function. J. Steroid Biochem. Molec. Biol. 40, 735-753.

Porterfield, S.P., Hendry, L.B. (1998): Impact of PCBs on thyroid hormone directed brain development. Toxicol. Indust. Health 14, 103-120.

Pugeat, M.M., Dunn, J.F., Nisula, B.C. (1981): Transport of steroid hormones: Interaction of 70 drugs with testosterone-binding globulin and corticosteroid-binding globulin in human plasma. J. Clin. Endocrin. Metab. 53, 69-75.

Pugeat, M., Dechaud, H., Emptoz-Bonneton, A., Lejeune, H., Lecoq, A., Tourniaire, J., Forest, M.G. (1986): Steroid drugs and plasma steroid-binding protein (SBP, CBG) interactions, p. 397-414. In: Binding proteins of steroid hormones (Forest, M.G., Pugeat, M., eds.). John Libbey & Comp. Ltd., London.

Pugeat, M., Lejeune, H., Dechaud, H., Emptoz-Bonneton, A., Fleury, M.-H., Charrié, A., Tourniaire, J., Forest, M.G. (1987): Effects of drug administration on gonadotropins, sex steroid hormones and binding proteins in humans. Hormone Res. 28, 261-273.

Robbins, J. (1992): Thyroxine transport and the free hormone hypothesis. Endocrinology 131, 546-547.

Rosner, W., Khan, M.S., Romas, N.A., Hryb, D.J. (1986): Interactions of plasma steroid-binding proteins with cell membranes, p. 567-575. In: Binding proteins of steroid hormones (Forest, M.G., Pugeat, M., eds.). John Libbey & Comp. Ltd., London.

Rosner, W., Hryb, D.J., Kahn, M.S., Nakla, A.M., Romas, N.A. (1999): Sex hormone-binding globulin mediates steroid hormone signal transduction at the plasma membrane. J. Steroid Biochem. Molec. Biol. 69, 481-485.

Sandberg, A.A., Slaunwhite, W.R. (1959): Transcortin: A corticosteroid-binding protein of plasma. II. Levels in various conditions and the effects of estrogens. J. Clin. Invest. 38, 1290-1297.

Savitz, D.A., Arbuckle, T., Kaczor, D., Curtis, K.M. (1997): Male pesticide exposure and pregnancy outcome. Am. J. Epidemiol. 146, 1025-1036.

Selby, C. (1990): Sex hormone binding globulin: origin, function and clinical significance. Ann. Clin. Biochem. 27, 532-541.

Shelby, M.D., Newbold, R.R., Tully, D.B., Chae, K., Davis, V.L. (1998): Assessing environmental chemicals for estrogenicity using a combination of in vitro and in vivo assays. Environ. Health Perspect. 104, 1296-1300.

Sher, E.S., Xu, X.M., Adams, P.M., Craft, C.M., Stein, S.A. (1998): The effects of thyroid hormone level and action in developing brain: Are these targets for the actions of polychlorinated biphenyls and dioxines? Toxicol. Indust. Health 14, 121-158.

Siiteri, P.K., Murai, J.T., Hammond, G.L., Nisker, J.A., Raymoure, W.J., Kuhn, R.W. (1982): The serum transport of steroid hormones. Rec. Prog. Horm. Res. 38, 457-510.

Soto, A.M., Sonnenschein, C., Chung, K.L., Fernandez, M.F., Olea, N., Olea Serrano, F. (1995): The E-SCREEN Assay as a tool to identify estrogens: An update on estrogenic environmental pollutants. Environ. Health Perspect. 103 (Suppl. 7), 113-122.

Toppari, J., Larsen, J.C., Christiansen, P., Giwercman, A., Grandjean, P., Guillette, L.J. Jr., Jégou, B., Jensen, T.K., Jouannet, P., Keiding, N., Leffers, H., McLachlan, J.A., Meyer, O., Müller, J., Rajpert-De Meyts, E., Scheike, T., Sharpe, R., Sumpter, J., Skakkebaek, N.E. (1996): Male reproductive health and environmental xenoestrogens. Environ. Health Perspect. 104 (Suppl. 4), 741-803.

Vermeulen, A. (1986): TeBG and CBG as an index of endocrine function, p. 383-

396. In: Binding proteins of steroid hormones (Forest, M.G., Pugeat, M., eds.). John Libbey & Comp. Ltd., London.

Vidal-Puig, A.J., Munez Torres, M., Jodar Gimeno, E., Garcia, Calvente, C.J., Lardelli, P., Ruiz Requena, M.E., Escobar, Jimenez, F. (1994). Ketoconazole therapy: hormonal and clinical effects in non-tumoral hyperandrogenism. Eur. J. Endocrinol. 130, 333-338.

Wilson, J.D., Foster, D.W. (1992): Hormones and hormone action, Introduction, p. 1-8. In: Williams Textbook of Endocrinology (Wilson, J.D., Foster, D.W., eds.). W.B. Saunders Comp., Philadelphia.

Zacharewski, T. (1997): In vitro bioassays for assessing estrogenic substances. Environ. Sci. Technol. 31, 613-623.

7 BIOMONITOR SYSTEMS DERIVED FROM THYLA-KOIDS FOR THE DETECTION OF PHYTOTOXIC SUBSTANCES

Heide Schnabl[1] and Stefanie Trapmann[2]

[1]University of Bonn, Institute for Agricultural Botany, Department of Physiology and Biotechnology of Plants, Karlrobert-Kreiten-Straße 13, 53115 Bonn, Germany
[2]European Commission, Joint Research Centre, Institute for Reference Materials and Measurements, IRMM, Retieseweg, 2440 Geel, Belgium

Abstract. The detection of phytotoxic compounds, such as herbicides, environmental substances, polyphenols, xenobiotics and phytohormones, is often based on the inhibition of the photosynthetic electron transfer between photosystem-II and photosystem-I when chloroplast thylakoids from *Vicia faba*, freshly isolated or lyophilized, were used. The lyophilization of the biological material ensures long-term stability and allows continuous supply without permanent cooling for more than one year. Within a few min the inhibition of photosynthetic electron flow by phytotoxic compounds is measured fluorometrically. The increase of fluorescence induced by herbicides is shown to be correlated with a reduction in light-dependent oxygen consumption. The site of inhibition by herbicides or other toxic compounds is indicated as the QB-binding site of the D1 protein inhibiting the plastoquinone reduction. Another group of herbicides is targeted at the cytochrom-b6/f-complex blocking the plastohydroquinone-oxidation. A third group of phytotoxins is able to perform redox interactions with photosystem-I transferring the electrons to oxygen and leading to a consumption of oxygen.

Fluorescence measurements using thylakoids offer all requirements as a preliminary screening tool for phytotoxic contamination. The biosensor is a very effective early warning system because of its low detection limit of 0.2 - 0.3 µg/l and its ability to indicate every photosynthetic active substance.

7.1   Introduction

Analytical methods are too time and cost consuming for quantifying the wide spectrum of pesticides, phytotoxins, xenobiotics, phytohormones and other substances with biological effects. Bioassays, using biological units from plants, are therefore required to identify toxic effects on plants caused by various environmental compounds and their metabolites. Especially, thylakoid membranes of chloroplasts (Bausch-Weis et al. 1994), offer a fast and cost-effective prescreening test system for monitoring contaminants in different samples from water, air, soil and food. In order to avoid the time consuming process of growing plants and isolating thylakoids, a new lyophilization technique of thylakoids was investigated, delivering long-term stabile membranes with high vitality and activity. 87-99% activity relative to the activity directly after thylakoid isolation was conserved for more than one year (Zimmermann et al. 1996, 1999). The advantage of lyophilization is the tolerance of freeze-dried material towards temperature modifications. No permanent cooling chain is necessary for maintaining thylakoid vitality, which allows the continuous and reproducible supply of the photosynthetic active biological unit for each application of the biosensor.

Lyophilized thylakoid membranes from higher plants serve as fast, specific, sensitive and cost effective indicator for herbicide determinations. Within a few min the photosynthetic electron transport in the thylakoid membranes can be measured and photosynthetic active substances can be detected in soil and compost samples (Helfrich et al. 1998) as well as water samples, such as drinking water, waste and surface water (Trapmann et al. 1998, Trapmann et al. 2000, Schnabl et al. 2000).

The photosynthetic electron transfer (PET) is completely maintained by the isolated and lyophilized thylakoids, except for ferredoxin (Izawa 1980) which is washed out from the membranes during isolation. The physiological electron acceptor FMN (flavin mononucleotide), however, is able to take over electrons from photosystem-I (Bausch-Weis et al. 1994) restoring the electron transport chain. The PET is measured using two strategies, the pulse-amplitude-modulation chlorophyll fluorescence (PAM 2000, Trapmann et al. 1998, Trapmann et al. 2000; Schnabl et al. 2000) and the oxygen consumption induced by substances which perform redox interactions with photosystem-I (PS-I) and transfer electrons to molecular oxygen. The oxygen consumption is measured with the help of a specialised oxygen electrode (Biolytik, Bochum, Germany; Helfrich et al. 1998; Schnabl et al. 2000).

Thus, thylakoids provide information on effect parameters such as phytotoxicity due to the binding of phytotoxic substances derived from different sources of environmental contamination. Three anchoring sites in thylakoid membranes are known to bind bioeffective ligands (Hock et al. 1995). Examples for the different ligand types, such as PS-II and -I active herbicides as well as other phytotoxic compounds, are presented in relation to their physiological response.

7.2 Results

7.2.1 Effect of PS-II and PS-I herbicides on isolated lyophilized thylakoids: Qualitative evaluation

The labour-intensive cultivation of plants and the isolation of thylakoids restricted the application of the bioassay considerably. To overcome this problem a production and conservation strategy was developed by lyophilizing the thylakoids (Zimmermann et al. 1996, 1999). In brief, the protective lyophilization was achieved by a step-wise sublimation and desorption at temperatures which gradually increased from $-20°$ to $+20°$ C and decreasing pressure conditions (from 0.15 to 0.01 mbar) within 41 hours (Zimmermann et al. 1999). The advantage of the lyophilization can be seen in Table 7.1. In these experiments conditions are simulated which occur during thylakoid transport to application sites or shipping to the user when a preliminary thawing process is inevitably. The activities of lyophilized and non-lyophilized thylakoids are

156

compared after the fresh isolation and after a 6 months storage period (-22° C). An intermediate period at room temperature for 3 days was followed again by a freezing process (-22° C). The final activity of non-lyophilized thylakoids resulted in a 50% loss during exposure to room temperature. In contrast, the lyophilized thylakoids showed only a reduction of 9%.

Tab. 7.1: Comparison of the activity of non-lyophilized and lyophilized thylakoids after a three day exposure to room temperature (RT)

Thylakoids	Activity (%)				
	after Isolation	after 6 months (-22°)	after 3 days (RT)	after 3 days (-22°)	Final activity (%)
Lyophilized	0,64	0,61	0,57	0,58	91
Non lyophilized	0,64	0,63	0,35	0,32	49

Isolated thylakoids usually retain all components of photosynthetic electron transport except for ferredoxin (Izawa 1980), the physiological electron acceptor flavin mononucleotide (FMN) was found to be able to take over electrons from PS-I and to restore the electron flow in thylakoids (Bausch-Weis et al. 1994). On the basis of the restored (by means of FMN), artificial electron chain between electron donors and acceptors, the presence of herbicides can be measured differentially. Herbicides, which inhibit plastoquinone reduction (such as atrazine, simazine, diuron, isoproturon, bromacil) or plastohydroquinone oxidation (dinitroanilin, such as trifluralin) in PS-II block the electron chain and cause thereby an increase of fluorescence (Fig. 7.1). However, many other phytotoxic substances act in the same manner, such as allelochemicals or short-chain fatty acids (Tab. 7.4). Since approximately 50% of the commercially available herbicides inhibit the electron flow between PS-II and PS-I (Youngman and Elstner 1988) a reproducible and sensitive way for detecting the quantum of electron flow is required and delivered by the fluorometric method of the biosensor. In this case the buffer system has to be replaced by the electron acceptor FMN. However, herbicides, which perform redox interactions with PS-I (bipyridins, such as diquat and paraquat), accept electrons in the same way as FMN. In both cases electrons are transferred to oxygen leading to an oxygen consumption. These types of herbicides (of course not only herbicides but also other substances which act in the same way) can only be detected without addition of FMN (Fig. 7.1).

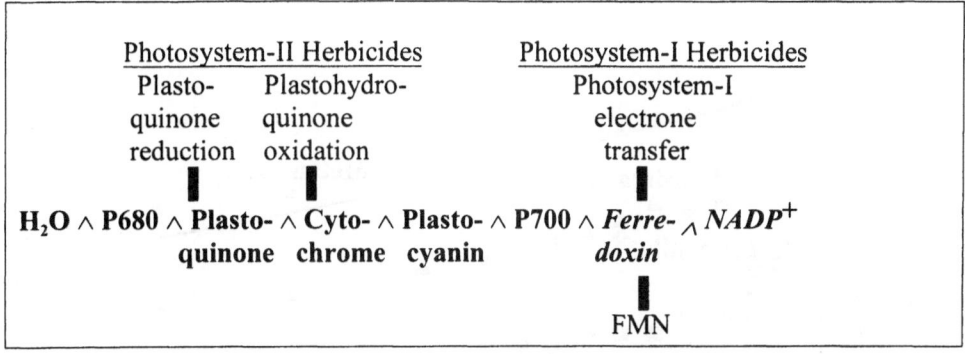

Fig. 7.1: Model of photosynthetic electron transport in isolated thylakoids with the sites of herbicide inhibition and the site of the artificial electron acceptor FMN. Components which can be leached out during the isolation process of thylakoids are shown in italic. In order to isolate thylakoids, *Vicia faba* L. plants were grown under standardised conditions in growth chambers in light (12 000 Lux, 24°C) and dark (21°C) rhythm of 9 and 15 h. The isolation of photoactive thylakoids from leaves of 3 week old plants was carried out according to Cohen and Baxter (1990). After chopping, filtration and centrifugation a hypotonic medium leads to osmotic rapture of all remaining chloroplasts. The activity of thylakoids was maintained by ascorbate (1%) which limits the production of oxygen radicals and the process of lipid peroxidation during thylakoid isolation. Addition of magnesium chloride (10 mmol) protects the central atom of chlorophyll from being leached out and sucrose (400 mmol) extends the durability of the biological units (Trapmann et al. 1998; Zimmermann et al. 1996)

7.2.2 Quantitative measurements based on fluorometry

The interactions between PS-II herbicides at different concentrations and their effects on the inhibition of electron flow are compared in Fig. 7.2. The individual inhibition

158

and the correlation between herbicide concentration and effect of different herbicides (metamitron, simazine, propanil, linuron, diuron), which are known as inhibitors of the electron transport at plastoquinone reduction site, are demonstrated.

The dose-effect correlation of every herbicide indicates a range in which the dose-effect curve is almost linear. In the case of diuron, this range is linear up to 16 µg/l, for simazine up to 30 µg/l, for linuron up to 15 µg/l, for propanil up to 24 µg/l, and for metamitron up to 100 µg/l. The gradient of the calibration line is a first hint for the sensitivity of the method.

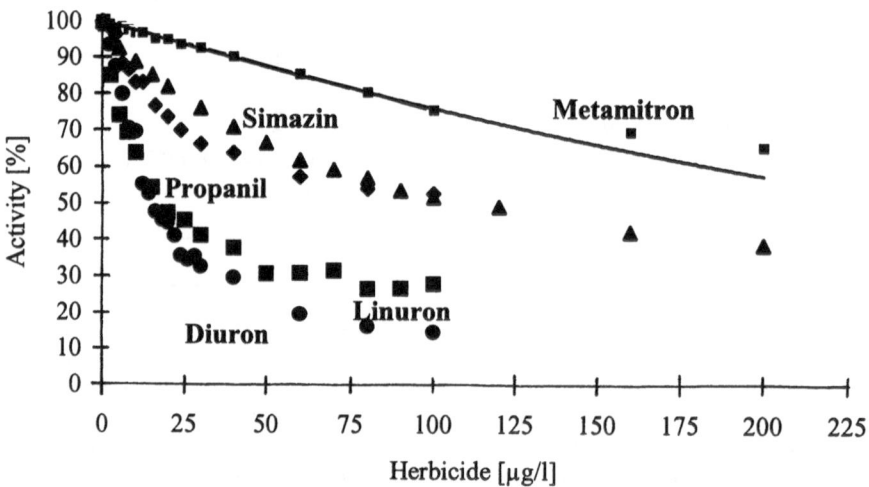

Fig. 7.2: Dose-effect correlations of different herbicides

In Table 7.2, the inhibition by different PS-II herbicides is given. Diuron, linuron, and ametryn proved to be most sensitive. In order to compare the inhibition effects of some herbicides the calculation of equivalents is required. Since the analysis of diuron is best-documented, diuron was chosen for calculating the equivalents of other herbicides, using the fluorometric method of blocking electron transfer in thylakoids.

Thus, the inhibition effect of 1 µg/l herbicide was compared with the inhibition of 1 µg/l diuron, the calibration lines were calculated for each herbicide. For the calculation of diuron equivalents it is important to analyse the type of herbicide interactions. A comparison between calculated and detected inhibition effects of herbicide mixtures showed that the PS-II active herbicides act in an additive way (Fig. 7.3). This is an important prerequisite for calculating the equivalents because the PET method is not able to distinguish between different herbicides and determines a cumulative parameter. In the same way, metabolites of herbicides are detected due to their inhibiting effect on the electron transfer (Trapmann et al. 1998).

In order to specify the effects of some other non-photosynthetic pesticides the PET inhibition was applied to trifluralin, an inhibitor of microtubuli, chlorpyriphos and terbufos, two organophosphates, and the insecticide carbaryl. As expected the gradient of the calibration lines are flat, since their effects are observed only in the ppm range. Fig. 7.4 shows the inhibition rate of the fluorescence with carbaryl as an example.

Tab. 7.2: Inhibition of different PS-II active herbicides and calculated diuron equivalents

Herbicide group	Herbicide	Inhibition of 1 µg/l herbicide	Diuron equivalents
Urea derivates	Diuron	4.3 %	1.0
	Linuron	3.7 %	0.9
	Methabenzthiazuron	1.4 %	0.3
	Isoproturon	1.8 %	0.4
Triazines	Metamitron	0.3 %	0.1
Chlortriazines	Atrazine	4.8 %	1.1
	Simazine	0.3 %	0.1
Methylthiotriazines	Ametryn	5.9 %	1.4
Biscarbamates	Phenmedipham	2.7 %	0.6
	Desmedipham	3.1 %	0.7
Anilides	Propanil	1.5 %	0.3
Uracils	Bromacil	1.0 %	0.2
Heterocycles	Bentazon	0.7 %	0.2
Nitriles	Ioxynil	< 0.1 %	< 0.1

160

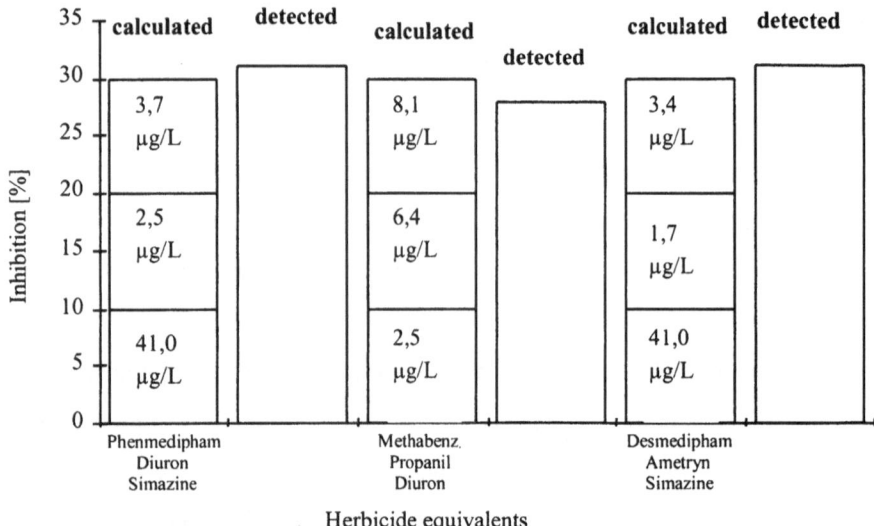

Fig. 7.3: Comparison of calculated and detected inhibition effects of herbicide mixtures in order to determine the type of interactions of PS-II herbicides. A dose equivalent corresponds to an inhibition effect of 5%; additive interactions show an inhibition effect of 30%

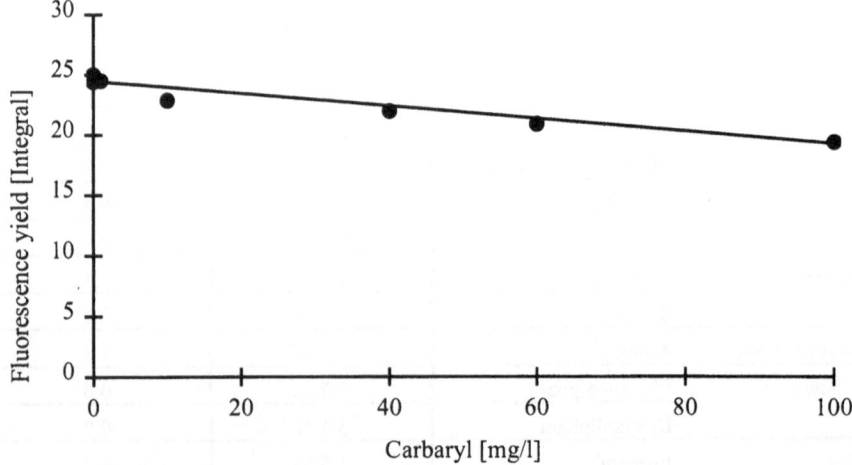

Fig. 7.4: The inhibition of photosynthetic electron-transfer in thylakoids by the carbamate insecticide carbaryl

No interference with the determination of herbicides was observed in the pH-range between 2 and 12 (data not shown; Trapmann et al. 2000). Furthermore, the calibration was not affected by concentrations of several inorganic ions, 10-times higher than the limits in the German drinking water guideline (Fig. 7.5). The only exceptions were calcium and sodium concentrations exceeding the limits of 400 and 150 mg/l (Trapmann et al. in press).

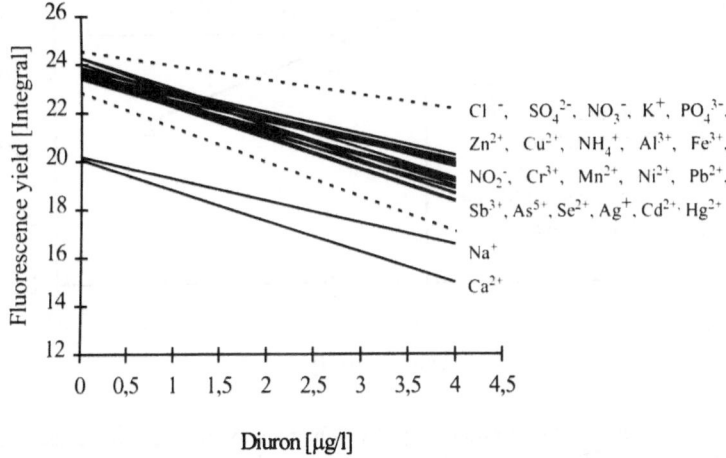

Fig. 7.5: The influence of various inorganic ions in concentrations 10-times higher than the limits in the German drinking water guideline on the sensitivity of lyophilized thylakoids for diuron. The confidence interval of reference measurements is based on 3-fold standard deviations (broken lines)

7.2.3   Measurements based on oxygen consumption

PS-II herbicides detected with FMN by PAM-fluorometry or oxygen electrode, respectively, are shown in Fig.7.6. Data of both techniques are compared using the herbicide diuron. In both cases linear dose-effect correlations are observed. The measurements of oxygen consumption show a higher standard deviation. Therefore, with fluorescence measurements the detection limit with diuron is 0.3 µg/l (the determination limit 0.7 µg/l). With oxygen measurements the limits are 10.7 and 33.4

µg/l, respectively. The differences in standard deviation may be due to the methods themselves. The fluorescence method is based on relative measurements, in contrast to the oxygen method, which determines absolute consumption.

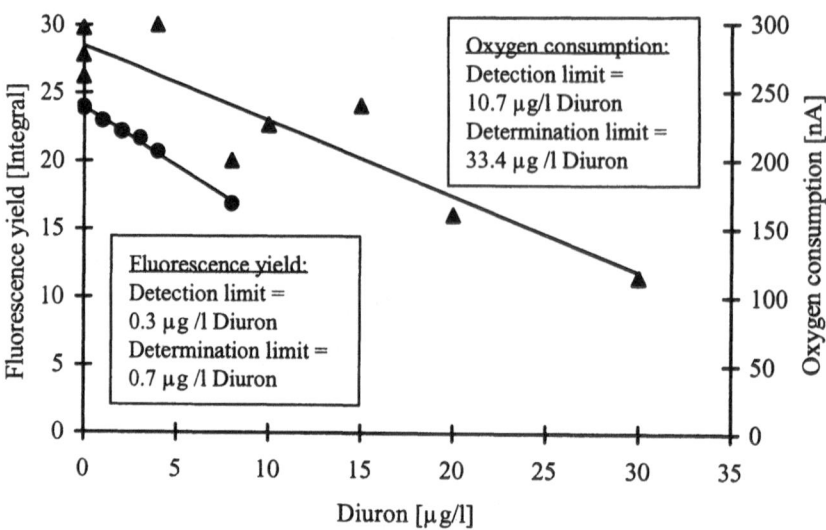

Fig. 7.6: The detection of diuron measured by PAM-fluorometry or by oxygen consumption in the presence of FMN, respectively. Using a pulse-amplitude-modulation chlorophyll fluorometer (PAM 2000, Walz Effeltrich, Germany) the electron transport was measured as a fluorescence yield after application of saturating light pulses (40 in 3s intervals; Bausch-Weis et al. 1994; Zimmermann et al. 1996; Trapmann et al. 1998; Helfrich et al. 1998). The fluorescence yield was used to determine the inhibition of the photosynthetic electron transport (PET): The lower the yield, the greater PET inhibition and the higher the amount of phytotoxic compounds inhibiting the electron transfer. The fluorescence yield was calculated as an empirical index for the quantum yield of electron flux (Schreiber and Bilger 1993). In order to measure the light dependent oxygen consumption of isolated thylakoids a specialised and miniaturised Clark electrode (Biolytik, Bochum) was used. Thylakoids were taken up in a buffer solution consisting of sucrose (320 mmol), HEPES (80 mmol), $MgCl_2$ (16 mmol), $CaCl_2$ (4mmol), pH 8.2 (Lindner et al. 1992; Overmeyer et al. 1994; Bausch-Weis et al. 1994; Helfrich et al. 1998)

7.2.4.   Photosystem-I active herbicides

PS-I photoactive herbicides can be detected only in absence of FMN with the oxygen electrode. The PS-I herbicides performing redox interactions with PS-I are able to accept electrons transferring them to molecular oxygen and thereby reducing it. The oxygen consumption measured in isolated thylakoids is determined using a specialised oxygen electrode containing a buffer system without FMN. Diquat is shown in Figure 7.7 as an example for the PS-I herbicides.

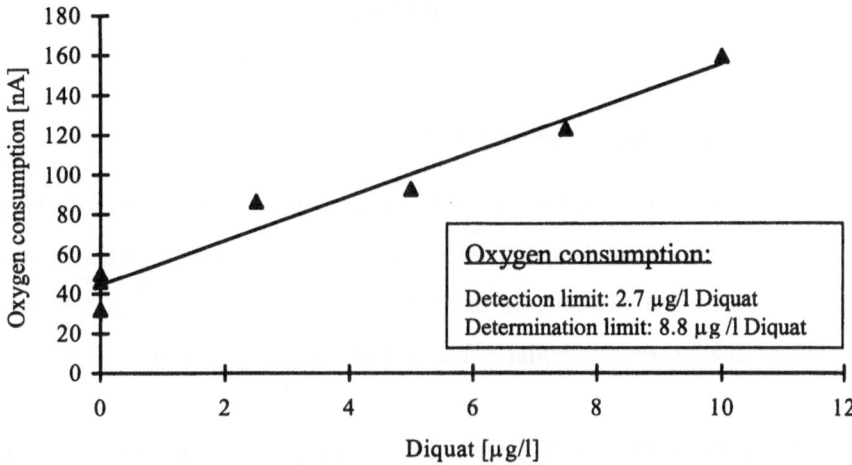

Fig. 7.7: Determination of the PS-I herbicide diquat with an oxygen electrode for measuring the oxygen consumption without artificial electron acceptor FMN

Tab. 7.3: Measurements of fluorescence and oxygen consumption with or without the artificial electron acceptor FMN in the presence of PS-I (here: diquat) and PS-II herbicides (here: diuron). The reference measurements are performed with distilled water

Herbicide	Measurements with FMN		Measurements with FMN	
	Fluorescence	Oxygen	Fluorescence	Oxygen
Reference	**23.8 (100 %)**	**242 (100 %)**	**6.8 (100 %)**	**34 (100 %)**
10 µg/l Diuron	16.3 (68 %)	158 (65 %)	6.1 (90 %)	31 (91 %)
10 µg/lDiquat	22.7 (95 %)	244 (101 %)	21.9 (320 %)	123 (362 %)
10 µg/l Diuron and 10 µg/l Diquat	15 (63 %)	145 (60 %)	10.7 (155 %)	28 (82 %)

164

Table 7.3 shows that measurements of fluorescence and oxygen consumption for PS-I herbicides (here: diquat) are possibly quantified wrongly by the presence of PS-II herbicides (here: diuron), when a mixture of PS-II and PS-I active herbicides is determined. In order to avoid errors in calculating contamination, a preceding determination of PS-II herbicide is necessary. The higher standard deviation of the oxygen method (Fig. 7.6) has to be accepted (Table 7.3).

7.2.5   Measurements of phytotoxic substances

We tested the thylakoid sensor with several other phytotoxic compounds, for example with allelochemicals, short-chain fatty acids, phytohormones and polyphenols. It was obvious that these components, if interacting with sites of electron transfer in thylakoid membranes, reveal similar effects as herbicides, which can be measured by

Table 7.4: The inhibition effects on PET (in %) induced by phytotoxic compounds (mM), such as cyclic hydroxamic acids (here BOA and APO, given in the text) and fatty acids, such as acetic and propionic acids

BOA	mM	0.5	1.0	1.5	2.0
	Inhib. %	2.0	2.5	3.0	4.5
APO	mM	0.01	0.05	0.1	0.2
	Inhib. %	4.0	21.0	25.1	28.2
Acetic acid	mM	150	200	250	300
	Inhib. %	22.3	37.3	63.0	68.9
Propionic acid	mM	100	150	200	300
	Inhib. %	6.0	14.5	31.2	73.0

the inhibition of the photosynthetic electron transfer. In Table 7.4 the inhibiting effects of cyclic hydroxamic acids are presented, using BOA (benzoxazolin-2-one) and its metabolite APO (2-amino-3H-phenoxazin-3-one), allelochemicals produced by root exudates of rye with phytotoxic properties. In the same table, examples for PET-inhibiting effects are given, which demonstrate the phytotoxic properties of products occuring during the ratting process of composts. Metabolisation during fermentation indicates the availability of mature compost products (Helfrich et al. 1998). However, at presence, the anchoring site of these phytotoxic substances in the electron chain is not well understood and further investigations are needed.

7.3   Discussion

Fluorescence measurements with thylakoid membranes as biological units meet all requirements as a preliminary screening tool for detecting herbicide contaminations. The biosensor is a very effective early warning system because of its low detection limit of 0.3 µg/l diuron and the ability to detect any photosynthetic acitive compound. Due to the additive interaction contaminating herbicides in samples can be calculated as cumulative parameter of all active compounds and can be expressed as diuron equivalents.

The thylakoid sensor indicates the presence of phytotoxic compounds with biological effects, such as herbicides, polyphenols, allelochemicals, xenobiotics, in a very fast, cost effective and specific way and with a low detection limit. The biosensor allows to monitor water samples - drinking water, waste, surface, rain water – as well as soil, compost and food samples. The inhibition of photosynthetic electron transfer of thylakoids by PS-II and PS-I active herbicides can be measured using the fluorescence yield. The high sensitivity level confirms the usefulness of the thylakoid sensor as a screening system for phytotoxic substances in drinking water without the necessity to perform preconcentration steps. The detection limit is higher when oxygen consumption is measured. When both physiological criteria of thylakoids, the inhibition of photosynthetic electron transfer and the oxygen consumption, are compared, the sensitivities differ by an order of magnitude, revealing the higher sensitivity of the fluorescence technique.

Since the long-term cultivation of plants in growth chambers and the lab-dependent isolation of thylakoids limit the application of the sensor in different fields, it was necessary to develop a technique for delivering a stabile membrane powder with high active properties. Lyophilizing thylakoids from higher plants offers a long-term stability for more than one year, without the necessity of a permanent cooling in a refrigerator. In this way, a monitoring in situ is possible.

Moreover, the isolated and lyophilized membrane system provides an option to differentiate phytotoxic substances because of their specialised binding capacities to different anchoring sites within the electron chain of PS-II and I. Due to the well investigated and characterised electron donors and acceptors it is possible to calculate the properties of anchored toxic substances on the basis of their physiological responses. The coupling of bioeffective substances (herbicides, allelochemicals, phytohormones), is restricted by the availability of specific target structures, for example the electron carriers exposing to these photosynthesis relevant ligands.

In the future, a concept for environmental analysis is planned based on the combination of bioeffective ligand binding and its chemical analysis after the extraction process. The success of the bioresponse-linked analysis depends on the choice of a suitable combination of target structures with substances indicating plant cell toxicity, in this case, via photosynthesis. These techniques open a new field in research and investigations for applying biosensors in practice and in research.

Acknowledgements. The authors are thankful to the DFG for financial support to H.S. (Trilateral project between Israel, Palestine and Germany). The Institute for Reference Materials and Measurements, Joint Research Centre, European Commission, Geel, Belgium as well as the Ministry of Science and Research. (MWF, Nordrhein-

Westfalen, Germany) are gratefully acknowledged for providing grants and research facilities to S.T.

7.4   References

Bausch-Weis, J., Overmeyer, S., Schnabl, H. (1994): Chloroplastenthylakoide als Herbiziddetektoren im Trinkwasser. Vom Wasser 83, 235-241.

Cohen, W.S., Baxter, D.R. (1990): Sulfhydryl-reagents and energy-linked reactions in monocot thylakoids. Plant Physiol. 93, 1005-1010.

Helfrich, P., Chefetz, B., Hadar, Y., Chen, Y., Schnabl, H. (1998): A novel method for determining phytotoxicity in composts. Compost Sci. & Util. 6(3), 6-13.

Hock, B., Fedtke, C., Schmidt, R.R. (1995): Herbizide-Entwicklung, Anwendung, Wirkungen, Nebenwirkungen. Georg Thieme Verlag Stuttgart, New York.

Izawa, S. (1980): Acceptors and donors for chloroplast electron transport. Methods in Enzymol. 69, 413-434.

Lindner, S.E., Overmeyer, S., Schnabl, H. (1992): Der Protoplastenbiotest - Ein Wirkungstest zur Herbiziddetektion in Gewässern. Angew. Bot. 66, 79-84.

Overmeyer, S., Wäber, M., Bausch-Weis, J., Peichl, L., Schnabl, H. (1994): Die Photosynthesehemmung pflanzlicher Protoplasten - Ein Indikator für Luftimmissionen. UWSF-Z. Umweltchem. Ökotox. 6(1), 5-8.

Schnabl, H., Helfrich, P., Trapmann, S. (2000): Thylakoids and protoplasts as toxicity monitoring systems, p. 177-184. In: New Microbiotests for Routine Toxicity Screening and Biomonitoring (Persoone, G., Jansson, C., De Coen, W., eds.). Plenum Publisher, London.

Schreiber, U., Bilger, W. (1993): Progress in chlorophyll fluorescence research: Major developments during the past years in retrospect, p. 151-173. In: Progress in Botany (Behmke, H.D., ed.). Springer Verlag, Berlin, Heidelberg.

Trapmann, S., Etxebarria, N., Schnabl, H., Grobecker, K.H. (1998): Progress in herbicide determination with the thylakoid bioassy. ESPR 5(1), 17-20.

Trapmann, S., Grobecker, K.H., Pauwels, J., Schnabl, H. (2000): Characterisation of lyophilized thylakoids as a biological unit to detect herbicides, p. 191-195. In: New Microbiotests for Routine Toxicity Screening and Biomonitoring (Persoone, G., Jansson, C., De Coen, W. eds.). Plenum Publisher, London.

Youngman, R.J., Elstner, E.F. (1988): Herbizide, p. 132-151. In: Schadwirkungen auf Pflanzen (Elstner, E.F., ed.). Wissenschaftsverlag, Mannheim, Wien, Zürich.

Zimmermann,G.M., Kramer, G.N., Schnabl, H. (1996): Lyophilization of thylakoids for improved handling in a bioassay. Environ. Toxicol. 15, 1461-1463.

Zimmermann, G.M., Trapmann, S., Pauwels, J., Schnabl, H (1999): Lyophilization of thyalkoids: A tool for long-term stability of the biological unit. Cryo-Letters 20, 229-239.

8 PROTEOMICS - A TOOL FOR BIORESPONSE-LINKED ANALYSIS

Ursula Bilitewski

National Research Centre for Biotechnology, Division Biochemical Engineering, Mascheroder Weg 1, D-38124 Braunschweig

Abstract. Cells respond to changes in their environment by modifications of proteins, either as primary or secondary reactions. These are effects on protein concentrations (up- or down-regulation of protein expression), enzyme activity or covalent protein modifications, such as phosphorylation or glycosylation. Driven by pharmaceutical research methods of protein analysis improved significantly leading to 2-D protein patterns with several hundred proteins describing the protein status of a cell under defined conditions. Investigations aiming at the construction of databases show correlations between incubations with defined drugs or ligands and influences on the protein pattern. Similar investigations with compounds of toxicological importance, which are present in the environment, show that they act on cells in a comparable way as pharmaceutical drugs. Thus, the establishment of a toxicological database should give new insights into the mode of action of environmental chemicals (structure-activity relationships), and allow the classification or even identification of unknown chemicals with respect to their mode of action.

8.1 Introduction

Cells respond to changes in their environment, e.g. changes in temperature, nutrient supply, presence of toxic or stimulating chemicals, by cell growth, cell death, cell migration, changes in metabolic pathways, in cell proliferation, cell activity (e.g. Mounho and Burchiel 1998, Kramer et al. 1987) or induction of apoptosis (e.g. Pryputniewicz et al. 1998), production of characteristic chemicals (e.g. Radice et al. 1998) or secondary metabolites, etc. Those reactions of cells are usually monitored by observations of the cells (growth or death) or the analysis of cell extracts or cell supernatants for characteristic compounds. These investigations may lead to the suggestion of molecular biomarkers for an observed effect (Timbrell 1998). For instance, the increased production of cell-associated interleukin-1α (IL-1) in murine keratinocytes was suggested as marker for skin allergens, whereas the increased release of IL-1 should indicate skin irritants, and the lack of an effect on IL-1 was attributed to chemicals having no effect on skin (Corsini et al. 1998). However, they do not allow to understand the cellular reactions in detail and do not give specific information on the mode of action of a chemical.

The functional molecules of cells are proteins (Williams 1999), and most chemicals, such as nutrients, drugs, toxicants, hormones, which show a specific biological activity, interact in the first instance with a protein. Relevant targets are transport proteins, either circulating in the blood (see also chapter 6) or being placed in cell membranes, receptors (see also chapter 4), transcription factors, enzymes, such as cholinesterases (see also chapters 5 and 2), kinases or phosphatases etc. From biochemical, cell and molecular biological investigations a number of related intracellular signal transduction cascades are understood in more or less detail (Devlin 1997, Krauss 1997, Heldin and Purton 1996), and they reveal that changes in the activity or concentration of one protein also lead to changes in other proteins, as a cell is a highly organized and regulated network of reactions. Reactions following the first interaction of a compound with a protein range from release of second messengers, activation/inactivation of enzymes, covalent modifications of proteins, formation of protein complexes, translocation of proteins to other cell compartments to the induction or repression of gene expression.

Considering this complexity of a biological systems, even of a defined cell kept in cell culture, approaches are nowadays favoured which allow the investigation of the system in its entirety. On the basis of the increased genomic information available from numerous sequencing programs, DNA-microarrays are suggested as a tool to explore how a particular chemical causes toxicity (Henry 1999). Based on expression analysis of a given cell type fingerprints for each chemical are expected. They allow on the one hand an identification of the mode of action of this chemical showing not only the expected first interaction but also possible side effects on the expression of other genes and on the other hand classifications of chemicals on the basis of similar fingerprints, i.e. similar modes of actions, which should allow in the future the identification of the relevant class also for unknown chemicals. On the basis of these ideas the Environmental Genome Project (EGP) was launched in the United States (Henry 1999, Weissenbach 1998) to understand the impact and interaction of environmental exposures on human disease.

However, the analysis of transcription pattern (which is named the "transcriptome") gives only a first information on the genes which are actually used from the whole genome, which represents the information bank. The final products of gene expression are the proteins, but there is no strict correlation between the abundance of expressed genes (mRNAs) and proteins found in a cell culture (Anderson and Seilhamer 1997, Anderson and Anderson 1998, Haynes et al. 1998). Reasons may be that mRNAs are still found for secreted proteins, which are not available for the cell anymore. Proteins controlling cellular metabolic processes are more frequently regulated at the translational level or by post-translational modifications. mRNAs differ in lifetimes among each other and, with respect to the corresponding protein, in that way that mRNAs for secreted proteins may have shorter half-lives than mRNAs for cellular enzymes. Consequently, the final understanding of cellular reactions has to include the analysis of the protein pool of cells, which is the final aim of the proteome research. Driven from investigations in pharmaceutical research, where the understanding of the mode of action of endogeneous ligands and of drugs is vital for the discovery of new molecular targets and the rational development of new pharmaceuticals, protein analysis made a tremendous progress within the last few years and developed into a major discipline in cellular research (Wilkins et al. 1997, Anderson and Anderson 1998, Lottspeich 1999).

In this chapter examples are given, which illustrate the complex changes on the protein level as already observed for some key biological ligands, such as Interferon gamma (IFN-γ), or growth factors, but also for toxicants, such as lead or dioxin. Comparisons of pattern of different compounds also showed already significant deviations, reflecting different molecular mechanisms. It is expected that drugs acting by similar mechanisms produce similar effects on the protein pattern. Then the pattern of protein changes observed after incubation with a sample / compound should provide sufficient information for classification and allow a new and more sophisticated approach to the study of structure-activity relationships (SARs) (Anderson and Anderson 1998).

8.2   Principles

Currently the most common used implementation of proteome analysis technology is based on the separation of proteins by two-dimensional gel electrophoresis (Wilkins et al. 1997, Anderson and Anderson 1998, Haynes et al. 1998, Lottspeich 1999). Usually proteins are first separated on the basis of their isoelectric points, using isoelectric focussing (IEF), and then in the second dimension on the basis of their molecular weights (polyacrylamide gel electrophoresis with SDS, SDS-PAGE). Traditionally the IEF separation was done in tube gels using carrier ampholytes (Lopez 1999), of which

6-18 can be cast and 1-15 run in parallel. Performance with respect to resolution and reproducibility was significantly improved by the introduction of immobilized pH gradients (IPG), which are now commercially available for various pH ranges and allow focussing of even alkaline proteins (Görg 1999). Using these strips a protocol was established for the combination with vertical or horizontal SDS electrophoresis (Görg and Weiss 1999). Several thousand proteins can be separated in this way using large-format gels. They can be visualized by organic dyes, with Coomassie Brilliant Blue as the most commonly used one, or by staining procedures using metal ions and colloids of gold, silver, etc. (Patton et al. 1999). These procedures lead to 2-D protein patterns. The position in the gel contains the information of the isoelectric point of the protein (pI) and of its mass. It may happen that proteins with similar properties are not well-separated and can not be distinguished (Wissing et al. 2000).

As proteins can not be selectively amplified by a reaction comparable to the polymerase chain reaction used for genes (PCR), the complexity of a protein mixture can only be reduced by a pre-separation of the total amount of proteins using suitable sample preparation procedures. They mainly include selective extraction procedures from the different cellular compartments, based on different solubilities of proteins (Rabilloud et al. 1997, Ramsby and Makowski 1999, Wissing et al. 2000). The complexity of the visible protein pattern can be reduced by selective staining of groups of proteins. This can be done by pulse-staining the cell culture with appropriate radioactive compounds such as ^{35}S-methionine/cysteine or $^{32/33}$P-orthophosphate or [γ-$^{32/33}$P]-ATP (Bizios 1999, Shaw et al. 1999a, b, Heim et al. 1999), but also by selective staining procedures of the already separated proteins, for example by antibodies against phosphorylated amino acids (Godova-Zimmermann et al. 1999), by metal chelates (Patton et al. 1999) or by specific chemical reactions for glycoproteins (Packer et al. 1999). However, these methods only give a picture of the protein pool and not yet the identity of proteins.

Therefore separation and staining of proteins is often followed by efforts leading to the identification of proteins. Depending on already available knowledge this is achieved by staining of proteins with antibodies specific for the protein under investigation after a blotting procedure. This allows identification of several isoforms or covalent modifications of the same protein. To date, however, digestion of proteins in the gel followed by matrix-assisted laser desorption/ionization-mass spectrometry (MALDI-MS) is the most often used procedure (Haynes et al. 1998, Shaw et al. 1999a, b, Witzmann et al. 1999, Godovac-Zimmermann et al. 1999). The comparison of the peptide pattern to information from protein databases leads to the identification at least of proteins, which were already identified by previous research. For completely unknown proteins classical methods, such as sequencing and functional assays, are required.

8.3   Applications

In the following some examples are given, in which influences on the protein pattern of cell cultures of either biological ligands or environmentally relevant chemicals were described.

8.3.1   Effects of IFN-γ on HeLa cells

IFN-γ is an important lymphokine, which is produced by T-lymphocytes and natural killer cells and involved in the regulation of antigen-specific and nonspecific immunological functions. It acts through binding to a specifc receptor on the target cells, which is activated upon ligand binding by tyrosine phosphorylation through kinases associated with the cytoplasmic domain of the IFN-γ-receptor. Subsequently further proteins are phosphorylated and activated leading finally to the activation of

IFN-γ-specific DNA regulatory elements. The number of known IFN-γ-regulated genes has increased to over 200 (Shaw et al. 1999a). HeLa cells were used as a model human cell line and treated with IFN-γ. Semiconfluent monolayers were pulse-labelled with ^{35}S-methionine/cysteine for 6 h after various intervals after addition of IFN-γ (up to 48 h) leading to kinetic information about changes in the protein pattern. In not-IFN-γ-treated cells 21 proteins were identified and mapped using a 2D PAGE-(IPG; pH: 4-7) system for protein separation and MALDI-MS and N-terminal sequencing for protein identification (Shaw et al. 1999b). Comparing the pattern from IFN-γ-treated cells to those from not-treated cells showed at least 8 proteins, the expression of which was altered due to the IFN-γ-treatment. Seven proteins were upregulated, and only 1, which proved to be proteasome subunit Y, was down-regulated. All 8 proteins showed an incubation time-dependent influence on their concentration. The strongest increase (22-fold) was observed for tryptophanyl-tRNA-synthetase after 24 h of IFN-γ-incubation. Three alternative forms of this enzyme were identified to be upregulated, however to different degrees. Significant increases were also observed for the expression of lipocortin (6 fold after 6 h), IGUPI-5111 (interferon-upregulated protein 5111, 6 fold after 24h) and cytokeratin 17 (3 fold after 12h). Most of these proteins were already known to be regulated in an IFN-γ-dependent manner. The data will form the basis of a 2D PAGE database accessible through the World Wide Web.

8.3.2   Lead exposure of rat kidneys

Lead is a potent neuro- and nephrotoxin in humans and a renal carcinogen in rats (Witzmann et al. 1999). Indicating regional renal physiologic differences between renal regions, histologic examination and enzyme assays confirm that the renal cortex and medulla are constitutively and biochemically different. Consequently, probing the effect of lead exposure on the expression of proteins in whole kidney homogenates may not reflect regiospecific differences in susceptibility to lead. Therefore Witzmann et al. (1999) compared protein pattern of kidney corticol and medullary cytosols from rats with and without injection of lead acetate. They resolved 727 protein spots in the cortex cytosol and 716 in the medulla cytosol using 20 x 25 cm 2D gels with IEF-tubes for the first dimension and SDS gradient gels for the second. Analysis was done first by Coomassie Blue staining, and in addition by Western blotting, immunoanalysis and by MALDI-MS, respectively. The abundance of 122 proteins differed significantly between both regions of the kidney. Approximately 30 proteins were unique to one of the regions. Lead exposure significantly altered the abundance of 78 proteins in the cortex and 16 proteins in the medulla, of which approximately equal numbers were increased and decreased. From this group 13 different proteins were identified, of which some existed in various alternative forms. It was found that lead administration changed the post-translational modification of glutathione S-transferase (GST) in the renal cortex in addition to an increased expression and activity. The observed induction of aflatoxin B_1 aldehyde reductase, aldose reductase, transketolase and also of GST may be explained by lead-related oxidative stress, but some of the genes are also known to respond to other xenobiotics, and their induction is not typical for lead, comparable to elevated expression of HSP90 (heat shock protein), which is a prominent cytosolic stress protein indicative also for metabolic and osmotic stress.

8.3.3   TCDD treatment of murine B cells

2,3,7,8-tetrachlorodibenzo-p-dioxin (TCDD) and related compounds have been shown to cause many toxic effects, of which the effects on the immune system are the most characteristic features. Despite extensive research and detailed investigations of the

TCDD receptor, the biochemical mechanisms which mediate the effects are still not clear (Pryputniewicz et al. 1998, Kramer et al. 1987). Kramer et al. (1987) focussed on influences on kinase activities, namely on endogenous phosporylated proteins, due to TCDD-incubation of murine B lymphocytes. Cells were prelabeled with $^{32}PO_4$ and then incubated with TCDD. Cytosolic proteins were separated by 1-D SDS PAGE and radioactivity was documented. Six proteins (MW approx. 12 - 65 kDa) were found showing TCDD dose- and incubation time-dependent stimulation of phosphorylation. For comparison they stimulated the cells also with phorbol 12-myristate, 13-acetate (PMA), which is a known activator of B cells, probably via stimulation of protein kinase C (PKC). Stimulation of phosphorylation was found for 5 of the 6 proteins also for PMA. However, the proteins were not identified, in particular the protein with MW 45.2 kDa, which was specifically phosphorylated after TCDD treatment, was not further characterized. Nevertheless, the results indicate that both TCDD and PMA activate kinases, which are either different enzymes with overlapping substrate specificities or belong at least to some extent to the same intracellular network. These early investigations were limited by the resolution power of the used electrophoresis system, but show already possible effects of environmental chemicals on specific groups of intracellular proteins.

8.3.4   Signalling of PDGF in mouse fibroblasts

PDGF (platelet-derived growth factor) is a mitogen involved in cell stimulation, for example during wound healing and bone growth, and interacts with its target cells via a membrane bound receptor, which belongs to the group of tyrosine-kinase receptors (Heldin and Purton 1996). These receptors share a common structure with an extracellular ligand binding domain, one transmembrane domain and a cytosolic tyrosine kinase domain. Upon ligand binding receptor molecules dimerize and autophosphorylate on tyrosine residues. In the following this phosphorylation is used for binding and activation of additional proteins, often related to further phosphorylation reactions on either tyrosine or serine / threonine. The investigation of the PDGF influence on the protein pattern of cells was done with mouse fibroblasts (Godovac-Zimmermann et al. 1999). Their total cellular proteins were separated by 2-D electrophoresis (IPG pH 4-7, 11.5% SDS gels 16 x 14 cm). Staining with Coomassie Blue or silver showed approximately 3000 proteins, which makes a comparison to existing 2D-maps of proteins a tremendous effort.

Therefore Godovac-Zimmermann et al. (1999) focussed on proteins phosphorylated on either tyrosine or serine, as phosphorylation of proteins is directly related to the mode of action of the stimulation via PDGF. The detection of this sub-group of proteins was achieved by anti-phosphotyrosine and anti-phosphoserine antibodies incubated on a blot of the cellular proteins. Antibody binding was visualized by an alkaline phosphatase-labelled secondary antibody and staining with suitable enzyme substrates. This procedure visualized still at least 260 proteins phosphorylated on tyrosine, of which 44 showed strong intensity changes as a function of time after stimulation with PDGF, and approximately 300 proteins phosphorylated on serine with 50 being influenced by stimulation with PDGF. Some of those proteins were already identified by in-gel digestion with trypsin, MALDI-TOF-MS (MALDI-time of flight-MS) and data base analysis. Proteins were found which were already known to be involved in the PDGF-signaling cascade (e.g. serine/threonine protein kinase akt, vimentin, protein tyrosine phosphatase SYP and others). Others were known from other signal transduction systems (proto-oncogene tyrosine kinase fgr, phosphotyrosine phosphatase PTP-2 and others). But also proteins occurred, which were not known to be involved in signal transduction (plexin-like protein).

For comparison cells were also stimulated with endothelin, another polypeptide hormone. It could be shown, that endothelin and PDGF influenced the phosphorylation of different proteins, with the changes observed for PDGF being more pronounced. This type of experiments showed that protein patterns due to ligand binding to a receptor followed by signal transduction are to a certain degree specific for the corresponding ligand. However, intracellular networks involved in signal tranduction are rather complex and show overlaps between different stimulating compounds.

8.4 Discussion

It was shown that modern techniques of protein analysis allow the separation and visualization of a huge number of proteins. However, particular eukaryotic cells may possess several thousand proteins with concentrations differing in several orders of magnitude. Usually, the most interesting proteins are the less abundant. Therefore, it seems to be necessary to reduce the number of proteins separated in each run and to describe the protein status of a cell type at a given moment by a set of protein pattern. Reduction of the complexity of the protein mixture can follow mainly two different rationals:

(i) Based on already known biochemical reactions following the treatment of cells with a substance, a corresponding group of proteins is selected. Examples are the investigations of phosphorylated proteins after treatment with TCDD or with PDGF. This usually leads only to a reduction of the complexity of the protein pattern, but not of the protein mixture, as not-phosphorylated proteins are present in the gel, but are not stained. This approach was called "functional proteomics" (Godovac-Zimmermann et al. 1999).

The same ideas can be utilized for a reduction of the complexity of the protein mixture, if selective staining is achieved by specific affinity reactions, e.g. staining of phosphorylated tyrosines with an anti-phosphotyrosine antibody. These affinity reagents can be coupled to a polymer, such as agarose, and used for immunoprecipitation or other types of affinity purification of the corresponding proteins. Eluted proteins can then be separated by electrophoresis (Mounho and Burchiel 1998).

(ii) Alternatively, the cell can be fractionated into different compartments, such as the membrane fraction, cytosol and nucleus. Proteins associated with each of these compartments can then be solubilized by specific methods and separated. For separation different pH-ranges in IEF and different SDS-gradients in the second dimension can then be used, leading to protein patterns comprising proteins only from a given compartment with a certain solubility, pI-range and range of molecular weights. Thus, the whole protein pool is split into fractions without predetermining the function of proteins.

It is known that the primary reaction of a substance (drug or toxicant) is followed by secondary reactions which may be the main cause of the therapeutic or toxic effect. Several compounds may lead to similar secondary effects in a cell, visualized by identical parts in the protein pattern. This leads to a classification of compounds according to their mode of actions and on the other hand to a classification of proteins which are found to be co-regulated by a number of compounds. Thus, proteins will be found, which are representative for the primary reactions within the cell, and others, which are influenced by secondary reactions, and representatives of those classes of proteins may develop into novel biomarkers.

At present the proteomic approach is followed mainly in the medical/ pharmaceutical area, as it is expected that the correlation between drugs and the resulting protein pattern of a cell type gives a new basis for structure-activity relationships.

However, it is already known that similar effects are also observed for chemicals of importance in the environment or in toxicology. However, these investigations are more sporadic and not yet as systematic as for pharmaceutical applications (Anderson and Anderson 1998). Probably, one reason is that those investigations can only be done with cell cultures. Thus, a cell line has to be chosen which is, ideally, representative for the (toxic) effects under investigation. For a number of environmental compounds, however, the target organ is not known. Moreover, due to the complexity of cellular reactions and the variability of protein pattern, especially in response to the lack of nutrients, temperature or osmosis-induced stress, it will be vital, at least in the beginning, to collect data using cell cultures with optimized conditions for the growth of the cells, which are incubated for defined periods of time with the sample or compound under investigation.

Prior to the analysis of unknown compounds a database has to be established collecting data of many compounds acting on different types of cells, e.g. on B-lymphocytes, T-lymphocytes, keratinocytes, kidney or liver cells. The compounds represented in the database should reflect a spectrum of different molecular modes of action and different degrees and types of toxicity. Then the database can be used to classify compounds present in samples under investigation on the basis of their mode of action on the cell line represented by the resulting protein pattern.

Careful analysis of the secondary reactions will lead to predictions of proteins acting as targets for the observed cellular reaction, which can then serve as receptors in affinity-based analytical set-ups. If the affinity between the target protein and analytes is high enough so that binding is maintained even after cell lysis and protein extraction, the position of the protein in the gel will be shifted compared to the free protein. Therefore, identification of proteins by mass spectrometry or Western blotting may directly highlight proteins with bound ligands without the necessity of protein purification. The corresponding bound ligands can also be identified by mass spectrometry as in other examples of bioresponse-linked analysis.

8.5 References

Anderson, L., Seilhamer, J. (1997): A comparison of selected mRNA and protein abundances in human liver. Electrophoresis 18, 533-537.

Anderson, N.L., Anderson, N.G. (1998): Proteome and proteomics: New technologies, new concepts, and new words. Electrophoresis 19, 1853-1861.

Bizios, N. (1999): Eukaryotic cell labeling and preparation for 2-D, p. 49-52. In: 2-D Proteome Analysis Protocols (Link, A.J., ed.). Humana Press, Totowa.

Corsini, E., Primavera, A., Marinovich, M., Galli, C.L. (1998): Selective induction of cell-associated interleukin-1α in murine kreatinocytes by chemical allergens. Toxicology 129, 193-200.

Devlin, T.M. (1997): Textbook of biochemistry with clinical correlations. Wiley & Sons, New York.

Godovac-Zimmermann, J., Soskic, V., Poznanovic, S., Brianza, F. (1999): Functional proteomics of signal transduction by membrane receptors. Electrophoresis 20, 952-961.

Görg, A. (1999): IPG-Dalt of very alkaline proteins, p. 197-210. In: 2-D Proteome Analysis Protocols (Link, A.J., ed.). Humana Press, Totowa.

Görg, A., Weiss, W. (1999): Analytical IPG-Dalt, p. 189-196. In: 2-D Proteome Analysis Protocols (Link, A.J. ed.). Humana Press, Totowa.

Haynes, P.A., Gygi, S.P., Figeys, D., Aebersold, R. (1998): Proteome analysis: Biological assay or data archive? Electrophoresis 19, 1862-1871.

Heim, S., Wissing, J., Tegge, W., Bilitewski, U., Flohe, L., Frank, R. (1999): Quantitative imaging of protein phosphorylation events in cellular proteomes. IBC_s 3rd Annual International Conference Functional Proteomics, Boston.

Heldin, C.-H., Purton, M. (1996): Signal transduction. Chapman & Hall, London.

Henry, C.M. (1999): DNA microarrays in toxicology. Anal. Chem. 71, 462A-464A.

Kramer, C.M., Johnson, K.W., Dooley, R.K., Holsapple, M.P. (1987): 2,3,7,8-Tetrachlorodibenzo-p-dioxin (TCDD) enhances antibody production and protein kinase activity in murine B cells, Biochem. Biophys. Res. Commun. 145, 25-33.

Krauss, G. (1997): Biochemie der Regulation und Signaltransduktion. Wiley-VCH, Weinheim.

Lopez, M.F. (1999): 2-D Electrophoresis using carrier ampholytes in the first dimension (IEF), p. 111-128. In: 2-D Proteome Analysis Protocols (Link, A.J., ed.). Humana Press, Totowa.

Lottspeich, F. (1999): Proteomanalyse - ein Weg zur Funktionsanalyse von Proteinen. Angew. Chem. 111, 2630-2647.

Mounho, B.J., Burchiel, S.W. (1998): Alterations in human B cell calcium homeostasis by polycyclic aromatic hydrocarbons: Possible associations with cytochrome P450 metabolism and increased protein tyrosine phosphorylation. Toxicol. and Appl. Pharmacol. 149, 80-89.

Packer, N.H., Ball, M.S., Devine, P.L. (1999): Glycoprotein detection of 2-D separated proteins, p. 341 – 352. In: 2-D Proteome Analysis Protocols (Link, A.J., ed.). Humana Press, Totowa.

Patton, W.F., Lim, M.J., Shepro, D. (1999): Protein detection using reversible metal chelate stains, p. 331-340. In: 2-D Proteome Analysis Protocols (Link, A.J., ed.). Humana Press, Totowa.

Pryputniewicz, S.J., Nagarkatti, M., Nagarkatti, P.S. (1998): Differential induction of apoptosis in activated and resting T cells by 2,3,7,8-tetrachlorodibenzo-p-dioxin (TCDD) and its repercussion on T cell responsiveness. Toxicology 129, 211-226.

Rabilloud, T., Adessi, C., Giraudel, A., Lunardi, J. (1997): Improvement of the solubilization of proteins in two-dimensional electrophoresis with immobilized pH gradients. Electrophoresis 18, 307 – 316.

Radice, S., Marabini, L., Gervasoni, M., Ferraris, M., Chiesara, E. (1998): Adaptation to oxidative stress: Effects of vinclozolin and iprodione on the HepG2 cell line. Toxicology 129, 183-191.

Ramsby, M.L., Makowski, G.S. (1999): Differential detergent fractionation of eukaryotic cells, p. 53-66. In: 2-D Proteome Analysis Protocols (Link, A.J., ed.). Humana Press, Totowa.

Shaw, A.C., Larsen, M.R., Roepstorff, P., Justesen, J., Christiansen, G., Birkelund, S. (1999a): Mapping and identification of interferon gamma-regulated HeLa cell proteins separated by immobilized pH-gradient two-dimensional gel electrophoresis. Electrophoresis 20, 984-993.

Shaw, A.C., Larsen, M.R., Roepstorff, P., Holm, A., Christiansen, G., Birkelund, S. (1999b): Mapping and identification of HeLa cell proteins separated by immobilized pH-gradient two-dimensional gel electrophoresis and construction of a two-dimensional polyacrylamide gel electrophoresis database. Electrophoresis 20, 977-983.

Timbrell, J.A. (1998): Biomarkers in Toxicology. Toxicology 129, 1-12.

Weissenbach, J. (1998): Human genome mapping and sequencing: perspectives for toxicology. Toxicol. Lett. 102-103, 1-4.

Wilkins, M.R., Williams, K.L., Appel, R.D., Hochstrasser, D.F. (1997): Proteome research. Springer-Verlag, Berlin.

Williams, K.L. (1999): Genomes and proteomes: Towards a multidimensional view of biology. Electrophoresis 20, 678-688.

Wissing, J., Heim, S., Flohe, L., Bilitewski, U., Frank. R. (2000): Enrichment of hydrophobic proteins via Triton X-114 phase partitioning and hydroxyapatite column chromatography for mass spectrometry. Electrophoresis, submitted.

Witzmann, F.A., Grant, R.A., Wright, L.S., Kornguth, S.E., Siegel, F.L. (1999):

Regional protein alterations in rat kidneys induced by lead exposure. Electrophoresis <u>20</u>, 943-951.

Conclusions and Outlook

B. Hock and F. Scheller

The target analytes in environmetal analysis are typically present in the submicromolar range. Biological recognition elements providing the required affinity usually suffer from cross-rectivities. Thus species-specific analysis based on biomolecular recognition is in most cases not possible. Bioresponse-linked instrumental analysis combines biomolecular recognition, initiating a biological effect in the organism, and chemical analysis. Although this technology is only at its beginning, the power of this approach is obvious. It provides both, information on bioeffects with respect to the applied target structures as well as the chemical structure and concentration of active substances present in a sample.

Representative examples have been given in this book for the successful exploitation of biomolecular recognition principles for binding assays, which provide data on toxicity equivalents. Binding studies with the estrogen receptor, SHBG, chloroplast thylakoids as well as enzyme inhibition assays have been used to illustrate this principle. As the affinity of the analyte to its respective binding protein is related to the strength of the biological effect, sum parameters can be defined. The technology underlying these tests is equivalent to antibody binding assays as they are used in immunoanalysis, e.g. immunoassays. Whereas antibody binding to a specific target is not related to any biological effect of this target (the immune system is designed to discriminate between non-self and self), binding to a receptor triggers or prevents a biological effect. Enzyme inhibition assays, such as the acetylcholinesterase inhibition assays, follow the same concept if an inhibitor is bound to the enzyme and blocks its function leading to the disruption of a biological response. However, this binding is not determined by a binding assay but more directly by the decrease of the enzyme activity.

Nevertheless, these approaches cannot provide information on the chemical structure of the bioeffective substances. Therefore, bioeffective binding must be linked to chemical analysis. This can be achieved in different ways: (1) A screening for relevant (i.e. polluted) samples among large numbers of samples can be carried out by a suitable assay based on the target of consideration, e.g. binding assays with the proper binder (e.g. a receptor). If a positive signal is obtained, affinity chromatography with the biorecognition element is performed. The eluted bioeffective ligands can be subsequently identified by chemical analysis. (2) Fractionation of samples is followed by activity or binding tests with biological target molecules. Fractions with bioeffective substances are then further analyzed.

Advances in analytical instrumentation of biological macromolecules have triggered the concept of hyphenated technologies, which enable the automated coupling of binding assays with chemical analysis. Several variants of tighter coupling the biomolecular and chemical part appear to be feasible; it is expected that they will be realized in the near future. The most straightforward approach is the determination of the ratio of bound and free target structures (b/f) such as the binding sites of receptors or other binding proteins followed by the chemical analysis of bound analytes. Information on the strength of biological effect is based on the ratio between the bound and free target structure (b/f). If the sample contains several analytes that are bound to the target, equivalent concentrations being related to the biological response are measured. In contrast to classical binding assays, no tracers are required. Furthermore, matrix effects are minimized or eliminated since interfering substances either are separated from the target molecules or do not apply because of the omission of tracers.

Information on the chemical structure of the analyte bound to the target structure is provided by chemical analysis. Substances cross-reacting with the target structures are also detected by chemical analysis. Therefore, cumbersome array structures as they have been applied in immunoanalysis followed by chemometric evaluation are not required. Quasi-continuous analyses are feasible. The proper application of this approach reduces chemical analysis to those samples or fractions of samples that contain bioeffective compounds.

However, the scope of this strategy is restricted by the availability of representative target structures. For instance, DNA molecules only apply to DNA-relevant ligands, estrogen receptors only to certain classes of endocrine disruptors and so on. Therefore, the effectiveness of this type of analytics depends on the proper choice of biological target structures. It can be foreseen that this problem will be solved by a suitable combination of several different target structures and it may be necessary to combine specific targets (such as receptors) with more integral targets (such as some kinases or phosphatases), which may be realised by modular test batteries. The success of this concept depends on the optimal combination of the individual modules that can be exchanged according to the specific requirements. Presently, modules representing cell toxicity, genotoxicity, neurotoxicity and immunotoxicity appear to be most desirable.

It is obvious that the applicability of bioresponse-linked instrumental analysis depends to a great deal on the availability of relevant biochemical target structures. The use of proteomics is expected to considerably extend the availability of suitable targets. Recombinant approaches for their biosynthesis appear to be indispensable and will be provided by biotechnology.

Subject Index

194